Drugs from Marine Sources

Drugs from Marine Sources

Editors

Genoveffa Nuzzo
Carmela Gallo

MDPI • Basel • Beijing • Wuhan • Barcelona • Belgrade • Manchester • Tokyo • Cluj • Tianjin

Editors
Genoveffa Nuzzo
Institute of Biomolecular Chemistry—CNR
Italy

Carmela Gallo
Institute of Biomolecular Chemistry
Italy

Editorial Office
MDPI
St. Alban-Anlage 66
4052 Basel, Switzerland

This is a reprint of articles from the Special Issue published online in the open access journal *Applied Sciences* (ISSN 2076-3417) (available at: https://www.mdpi.com/journal/applsci/special_issues/Drug_Discovery_Marine_Sources).

For citation purposes, cite each article independently as indicated on the article page online and as indicated below:

LastName, A.A.; LastName, B.B.; LastName, C.C. Article Title. *Journal Name* **Year**, *Volume Number*, Page Range.

ISBN 978-3-0365-3367-4 (Hbk)
ISBN 978-3-0365-3368-1 (PDF)

Cover image courtesy of Dr. Guido Villani

Contents

About the Editors

Genoveffa Nuzzo is a full researcher at the Institute of Biomolecular Chemistry of the National Research Council (ICB–CNR). She graduated with a degree in Pharmaceutical Chemistry and Technology from the University of Naples in July 2008 with 110 cum laude with honorable mention and produced an experimental thesis on organic chemistry. She earned her PhD in Pharmaceutical Sciences from the University of Salerno, focusing her research on the exploration of the chemical diversity of marine organisms in search of "biomolecules for health". During her doctoral course, she spent time in Greece at the University of Athens. After completing her doctorate, she began working at the Institute of Biomolecular Chemistry (ICB–CNR), and her research activity is focused on the characterization of biomolecules from marine organisms, working on several projects aimed at identifying active molecules from marine sources. She is also a member of a Spin-Off BioSearch S.r.l. (a non-profit organization), whose goal is to discover new natural compounds from marine organisms for nutraceutical and pharmaceutical purposes.

Carmela Gallo is a full researcher at the Institute of Biomolecular Chemistry of the National Research Council (ICB–CNR). She graduated with a degree in Biology from the University of Naples in 2011 with 110 cum laude. She obtained her PhD title in Applied Biology from the University of Naples in 2015, where her research activity was focused on the identification of bioactive molecules from marine organisms playing an ecophysiological role and specific biochemical properties. She has continued her research at the ICB since 2015, specializing in the field of cell biology. As of present, she has developed a bioassay platform for human cell lines of myeloid origin for the evaluation of the immunomodulating properties of the natural compounds isolated from marine organisms and synthetic derivatives, with the aim of developing products for the formulation of vaccines and new principles for the treatment of pathologies related to the decompensation of the innate immune system.

Editorial

Drugs from Marine Sources

Carmela Gallo and Genoveffa Nuzzo *

Bio-Organic Chemistry Unit, Institute of Biomolecular Chemistry, Consiglio Nazionale delle Ricerche, Via Campi Flegrei 34, 80078 Pozzuoli, Italy; carmen.gallo@icb.cnr.it
* Correspondence: nuzzo.genoveffa@icb.cnr.it

Citation: Gallo, C.; Nuzzo, G. Drugs from Marine Sources. *Appl. Sci.* **2021**, *11*, 12115. https://doi.org/10.3390/app112412115

Received: 9 December 2021
Accepted: 14 December 2021
Published: 20 December 2021

Throughout history, natural products have afforded a rich source of compounds that have found many applications in the fields of pharmacology. In this context, Marine Natural Products (MNPs) represent a unique rich source of new metabolites, with diverse chemical structures, and various bioactivities and molecular characteristics adapted to specific interactions with cellular targets. The huge diversity and potential bioactivity of marine metabolites can serve as "lead" compounds for the discovery of modern drugs. Thus, these products are attractive targets because they could be used to overcome the challenge of treating a high number of diseases that affect human health.

This Special Issue has covered all fields of drugs research from marine natural products, including their isolation and characterization, in vitro and/or in vivo study of their biological activities, medicinal applications as well as synthetic approaches and related analogues.

In detail, two studies are reported in this Special Issue on the chemical characterization and biological activity of alginate from two species of *Sargassum*; in particular, alginate was investigated for wound-healing and antioxidant properties on a diabetic mouse model [1,2]. A third work reports the bioactivity-guided study of *Aquimarina* sp. with the identification of diketopiperazines and N-Phenethylacetamide and their effect on TGF-β-induced epithelial–mesenchymal transition [3].

Pharmacokinetic and drug metabolism tests are mandatory for expediting the progress of compounds with promising properties from discovery to development phase. For this purpose, a further manuscript is included about the investigation of a robust and fast analytical method for the measurement of Sulfavant A, a prototype for a new class of molecular adjuvants inspired by natural α-sulfoquinovosyl-diacylglycerols (α-SQDGs) occurring in diatoms by Ultra High Performance Liquid Chromatography coupled with High Resolution Mass Spectrometry (UHPLC-HRMS) [4].

As part of this Special Issue, two reviews about medicinal applications of natural products from marine sources have been reported. In particular a minireview details the potential applications of diatom-derived silica for the biomedical field [5]; finally, a review describes the advances in purpurin 18 research as a promising anticancer drug [6].

Author Contributions: Conceptualization, G.N.; review papers and writing editorial, C.G. and G.N. All authors have read and agreed to the published version of the manuscript.

Funding: This work was supported by the project "Antitumor Drugs and Vaccines from the Sea (ADViSE)" project (CUP B43D18000240007–SURF 17061BP000000011) funded by POR Campania FESR 2014–2020 "Technology Platformfor Therapeutic Strategies against Cancer"-Action 1.1.2 and 1.2.2.

Conflicts of Interest: The authors declare no conflict of interest.

References

1. Pudjiastuti, P. Characterization of Alginate from Sargassum duplicatum and the Antioxidant Effect of Alginate–Okra Fruit Extracts Combination for Wound Healing on Diabetic Mice. *Appl. Sci.* **2020**, *10*, 6082.
2. Wulandari, P.A.C.; Ilmi, Z.N.; Husen, S.A.; Winarni, D.; Alamsjah, M.A.; Awang, K.; Vastano, M.; Pellis, A.; MacQuarrie, D.; Pudjiastuti, P. Wound Healing and Antioxidant Evaluations of Alginate from Sargassum ilicifolium and Mangosteen Rind Combination Extracts on Diabetic Mice Model. *Appl. Sci.* **2021**, *11*, 4651. [CrossRef]
3. Lee, M.J.; Kim, G.J.; Shin, M.-S.; Moon, J.; Kim, S.; Nam, J.-W.; Kang, K.S.; Choi, H. Chemical Investigation of Diketopiperazines and N-Phenethylacetamide Isolated from Aquimarina sp. MC085 and Their Effect on TGF-β-Induced Epithelial–Mesenchymal Transition. *Appl. Sci.* **2021**, *11*, 8866. [CrossRef]
4. Nuzzo, G.; Manzo, E.; Ziaco, M.; Fioretto, L.; Campos, A.M.; Gallo, C.; D'Ippolito, G.; Fontana, A. UHPLC-MS Method for the Analysis of the Molecular Adjuvant Sulfavant A. *Appl. Sci.* **2021**, *11*, 1451. [CrossRef]
5. Sardo, A.; Orefice, I.; Balzano, S.; Barra, L.; Romano, G. Mini-Review: Potential of Diatom-Derived Silica for Biomedical Applications. *Appl. Sci.* **2021**, *11*, 4533. [CrossRef]
6. Pavlíčková, V.; Škubník, J.; Jurášek, M.; Rimpelová, S. Advances in Purpurin 18 Research: On Cancer Therapy. *Appl. Sci.* **2021**, *11*, 2254. [CrossRef]

applied sciences

Article

UHPLC-MS Method for the Analysis of the Molecular Adjuvant Sulfavant A

Genoveffa Nuzzo [1,*,†], Emiliano Manzo [1,†], Marcello Ziaco [2], Laura Fioretto [1], Ana Margarida Campos [1,3], Carmela Gallo [1], Giuliana d'Ippolito [1] and Angelo Fontana [1,4]

1 Bio-Organic Chemistry Unit, National Research Council-Institute of Bio-Molecular Chemistry (ICB-CNR), Via Campi Flegrei 34, 80078 Pozzuoli, Italy; emanzo@icb.cnr.it (E.M.); l.fioretto@icb.cnr.it (L.F.); am.campos@ibp.cnr.it (A.M.C.); carmen.gallo@icb.cnr.it (C.G.); gdippolito@icb.cnr.it (G.d.); afontana@icb.cnr.it (A.F.)
2 BioSearch Srl., Villa Comunale c/o Stazione Zoologica "A.Dohrn", 80121 Napoli, Italy; m.ziaco@icb.cnr.it
3 Institute of Protein Biochemistry, National Research Council, Via Pietro Castellino 111, 80131 Naples, Italy
4 Department of Biology, University of Napoli "Federico II", Via Cupa Nuova Cinthia 21, 80126 Napoli, Italy
* Correspondence: nuzzo.genoveffa@icb.cnr.it
† These authors contributed equally to this work.

Abstract: A fast and sensitive method that is based on Ultra High Performance Liquid Chromatography coupled with High Resolution Mass Spectrometry (UHPLC-HRMS) for the measurement of Sulfavant A, a molecular adjuvant with a sulfolipid skeleton, is described. The method has been validated over the linearity range of 2.5–2000 ngmL^{-1} using a deuterated derivative (d$_{70}$-Sulfavant A) as internal standard. Chromatographic separation is based on a UHPLC Kinetex® 2.6 µm PS C18 column and a gradient of methanol in 0.32 mM ammonium hydroxide solution buffered at pH 8. The lowest limit of quantification of Sulfavant A was 6.5 ngmL^{-1}. The analytical procedure was tested on an extract of mice lung spiked with 30, 300, and 1500 ng of Sulfavant A. The analysis revealed a precision and accuracy value (as a mean value of all the quality control samples analyzed) of 4.7% and 96% in MeOH and 6.4% and 93.4% in the lung extracts, respectively.

Keywords: sulfavant; sulfoquinovosyldiacylglycerols; sulfoglycolipids; mass spectrometry; UHPLC-MS; lipids

Citation: Nuzzo, G.; Manzo, E.; Ziaco, M.; Fioretto, L.; Campos, A.M.; Gallo, C.; d'Ippolito, G.; Fontana, A. UHPLC-MS Method for the Analysis of the Molecular Adjuvant Sulfavant A. *Appl. Sci.* **2021**, *11*, 1451. https://doi.org/10.3390/app11041451

Academic Editor: Magdalena Biesaga
Received: 7 January 2021
Accepted: 2 February 2021
Published: 5 February 2021

1. Introduction

Sulfur-containing lipids and, in particular, sulfoquinovosyl-diacylglycerols (SQDGs) have been proposed as factors in inflammation, immunity and infection. However, despite their dissemination, fast and accurate ultra-performance liquid chromatography–mass spectrometry (UPLC-MS) or ultra high-performance liquid chromatography coupled with mass spectrometry (UHPLC-MS) methods for their quantification are still scarce and mostly focused on sulphate sterols [1–4] and sulfatides [5–7]. Most of the lipid analysis, also containing sulfolipids, as reported in the literature, are based on a shotgun lipidomic approach [2,3] or traditional reverse phase (RP) - HPLC columns usually associated with complex mixtures of solvents used as eluents in order to obtain a good quality chromatogram [8].

In our ongoing drug discovery exploration for new active metabolites, we recently reported the immunomodulatory activity of 1,2-*O*-distearoyl-3-*O*-β-D-sulfoquinovosylglycerol, named Sulfavant A (**1**) (SULF A; Figure 1). SULF A is a synthetic glycolipid that is featured by a sugar unit of sulfoquinovose. The molecule is the prototype for a new class of molecular adjuvants inspired by natural α-sulfoquinovosyl-diacylglycerols (α-SQDGs) occurring in photosynthetic organisms [9]. SULF A (**1**) triggers in vitro maturation of dendritic cells, the master control of innate immune response, and in in vivo antigen-specific immunization [10–15]. The initiation of a systemic immune response by the stimulation of innate immune cells correlates to adjuvanticity and Pattern Recognition Receptors (PRR) - mediated signaling. SULF A is under preclinical trials as a vaccine adjuvant and its efficacy has been already proven in a murine model of vaccine against melanoma [9]. Interestingly, the product is not cytotoxic, but treated mice do not show progress of the tumour for more than 10 days after subcutaneous injection of B16F10 melanoma cells.

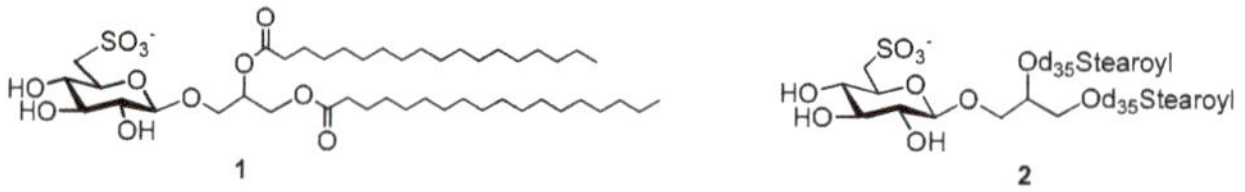

Figure 1. Chemical structure of Sulfavant A (**1**) deuterated Sulfavant A (**2**).

In the present study, we seek a robust and fast analytical method for the quantification of SULF A in biofluids and tissues during preclinical studies. To this aim, we report the implementation of a novel Ultra High-Performance Liquid Chromatography-High Resolution Mass Spectrometry (UHPLC-HRMS) method, together with the use of deuterated Sulfavant A (d_{70}-SULF A) (**2**) as internal standard (IS).

2. Materials and Methods

2.1. Chemicals and Reagents

SULF A (**1**) was prepared, as reported by Manzo et al. [9,10] and purified by HPLC before the analysis. The compound was eluted from a Phenomenex Phenl-Hexyl column (250 × 10 mm, 5 μm) in isocratic conditions using MeOH/H_2O 87:13 for 20 min. and the solvent was removed by evaporation under reduced pressure. Chromatographic purification of synthetic lipids was carried out by a JASCO system (PU-2089 Plus quaternary gradient pump) that was equipped with a MD-2018 Plus photodiode array detector (JASCO Europe Srl, Cremella, Italy) and Sedex 85 high-sensitivity LT-ELS detector (SEDERE, Alfortville, Paris, France). The purity of **1** was verified by ^{1}H NMR and LCMS.

UHPLC–HRESIMS analysis was performed on Q-Exactive hybrid quadrupole-orbitrap mass spectrometer (Thermo Scientific, Waltham, MA, USA) that was equipped with an Infinity 1290 UHPLC System (Agilent Technologies, Santa Clara, CA, USA). Methanol, water, and 25% ammonium hydroxide solution were all LC/MS grade. The chemical solvent and reagents were purchased from Merck Life Science S.r.l. (Milano, Italy).

2.2. Synthetic Strategy of d_{70}-Sulfavant A

The synthesis of d_{70}-SULF A was achieved by the ameliorated synthetic procedure for SULF A of Manzo et al. [10], which was adapted to the preparation of the analog totally deuterated on acyl portions; in detail, as compared to the literature procedure, deuterated stearic acid was used for the acylation of the 2′ 3′ 4′ 6′-*O*-tetracetyl-β-glucosyl-*R*/*S*-glycerol intermediate (Figure 2) [10,14,15]. Similarly to SULF A, compound **2** was purified by HPLC prior to use as an internal standard using the same experimental conditions.

2.3. Ultra-High-Performance Liquid Chromatography/High Resolution Mass Spectrometry (UHPLC/HRMS)

Chromatographic separations were achieved on UHPLC Kinetex® 2.6 μm PS C18 100 Å, LC Column 30 × 2.1 mm (Phenomenex, Italy), at 28 °C by a gradient elution of 0.32 mM ammonium hydroxide solution (0.005%), adjusted to pH 8.0 by acetic acid, and methanol (MeOH). Table 1 reports UHPLC gradient details. The flow rate was 0.5 mL min^{-1}. The injection volume was 4 μL, and the autosampler was maintained at 10 °C. MS analyses were carried out in electrospray ionization (ESI) negative mode with source parameters, as follows: spray voltage of 3.0 kV, capillary temperature of 320 °C, S-lens RF level of 60, sheath gas flow rate of 50, and auxiliary gas flow rate of 30. Full MS scans were acquired over the range of 150–1800 Da with a mass resolution of 70,000. The target value (Automatic Gain Control-AGC) was 1 × 10^6 and the maximum allowed accumulation time (IT) was 100 ms.

For the data-dependent MS/MS (ddMS2) analyses, the peaks of interest were selected and fragmented with a stepped normalized energy of 20–40 eV. AGC was 1×10^5 with IT 75 ms and 17, 500 mass resolution.

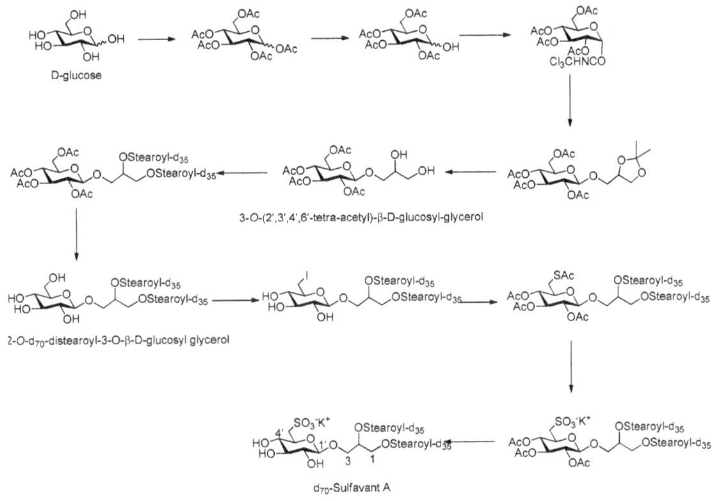

Figure 2. Chemical synthesis of deuterated Sulfavant A.

Table 1. Gradient program of the Ultra High Performance Liquid Chromatography (UHPLC) method.

Time (min)	Mobile Phase A (%) [a]	Mobile Phase B (%) [b]
0	85	15
1	85	15
6	98	2
8	100	0
10	100	0
11	85	15
15	85	15

[a] methanol [b] 0.32 mM ammonium hydroxide water solution (pH 8.0).

2.4. Standard Preparation and Stock Solution

SULF A and d_{70}-SULF A were dissolved in methanol to obtain stocks solutions at a concentration of 100 µg mL^{-1} stored at −20 °C; the dissolution was performed by sonicating for 30 min. at 30 °C. These two solutions were used to evaluate the linearity, calibration curve, limit of detection, and matrix effect. The evaluation of the above parameters is based on chromatograms of SULF A ($C_{45}H_{85}O_{12}S^-$; m/z 849.57672) and d_{70}-SULF A ($C_{45}D_{70}H_{15}O_{12}S^-$; m/z 920.01609) that were obtained as base peak extraction at m/z 849.576 and 920.016, respectively, with a mass tolerance of 5 ppm (Figure 3).

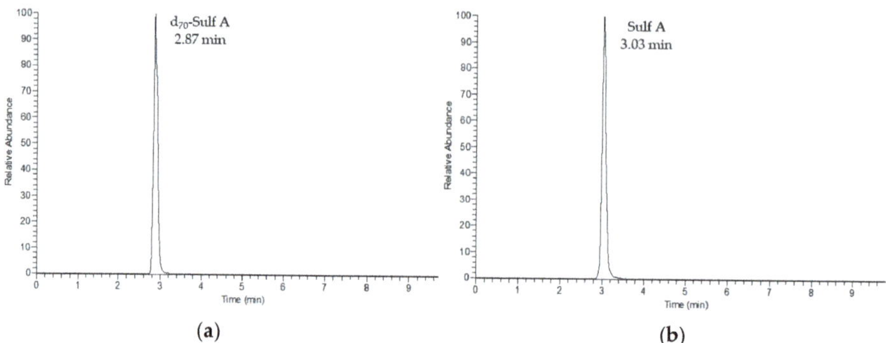

Figure 3. Chromatograms of d_{70}-Sulfavant A (**a**) and Sulfavant A (**b**) as base peak extraction at m/z 920.016 and 849.576 respectively, applying a mass tolerance of 5 ppm.

2.5. Calibration Curve, Linearity and Detection and Quantitation Limits (LOD and LOQ)

Before the analysis, the stock solution was sonicated for 30 min. at 30 °C The calibration curve was obtained by spiking increasing concentrations of standard solution at 1, 2.5, 5, 10, 50, 100, 500, 1000, and 2000 $ngmL^{-1}$. The average of three measurements was used for building up the calibration curve. The linearity of the calibration curve was also verified in the lipid extract matrix. Visual definition and the approach based on the "Standard Deviation of the Response and the Slope" (LOD = $3\sigma/S$; LOQ = $10\sigma/S$) was used in order to establish detection and quantitation limit [16].

2.6. Method Validation: Intra-and Inter-Day Accuracy and Precision

Three quality control (QC) samples of SULF A were prepared to validate the method: low-concentration (30 $ngmL^{-1}$), medium-concentration (300 $ngmL^{-1}$), and high-concentration (1500 $ngmL^{-1}$). Validation for intra-day and inter-day assay accuracy and precision was assessed on three independent days while using a triplicate of three QC. Each batch of runs comprised mobile phase blanks, calibrators (eight samples spanning the range of 2.5–2000 $ngmL^{-1}$), and quality control samples. Intra-day and inter-day accuracy and precision were determined by obtaining the mean concentration of the quality control samples from the calibration curve and the percent accuracy and coefficient of variation. The precision was expressed as a coefficient of variation [CV = (standard deviation/measured mean concentration) × 100], while the accuracy was expressed as [measured mean concentration/nominal concentration] × 100. The overall method accuracy and precision were determined by calculating the mean of accuracy and mean of the precision estimates, respectively, of all the quality control samples.

2.7. Evaluation of Matrix Effect on Mice Lung Extracts

The matrix effect (ME) was assessed on mice lung tissue post-extraction. Lipid extracts were achieved from five different frozen samples by homogenization of the tissues in MeOH (300 µL) at 12 °C for 2 min. and then following with MTBE extraction [17]. In detail, 1000 µL of MTBE were added to the samples and then left to mix for 5 min. Later, 250 µL of Milli-Q water was added and mixed again for 2 min. Subsequently, to induce the phase separation, the sample was centrifuged at 1000 g for 10 min. at 4 °C and the upper organic phase was recovered. Finally, the aqueous phase was re-extracted with MTBE, the upper phases combined, and the organic solvent removed under nitrogen stream. Each extract was weighted and resuspended in MeOH:CH_2Cl_2 (4:1) and then diluted with MeOH for LC/MS analysis.

Three different concentrations of SULF A (30, 300, and 1500 $ngmL^{-1}$) were spiked into 50 µg mL^{-1} of lipid extracts, together with 50 $ngmL^{-1}$ of IS (d_{70}-SULF A), in order to evaluate normalized matrix effect (nME). Each concentration was compared with the corresponding solution in pure MeOH. When considering peak area as instrument response, the percentage of nME was calculated, as reported in [18]:

$$nME\% = [(A_{solvent}/IS_{solvent})/(A_{matrix}/IS_{matrix}) - 1] \times 100$$

in which $A_{solvent}$ = peak area of the analyte SULF A in MeOH; A_{matrix} = peak area of the analyte SULF A in the post-extraction spiked matrix; $IS_{solvent}$ = peak area of the Internal Standard d_{70}-SULF A in MeOH; and, IS_{matrix} = peak area of the Internal Standard d_{70}-SULF A in the post-extraction spiked matrix.

Moreover, standards solutions at 2.5, 5, 10, 50, 100, 500, 1000, and 2000 ngmL^{-1} were also prepared in matrix (50 µg mL^{-1} of mice lung lipid extract) in order to evaluate whether the matrix effect (ME) affected the slope of the calibration curve. The average of three measurements was used. Chromatograms of SULF A and d_{70}-SULF A (as in MeOH) were obtained as base peak extraction at m/z 849.576 and 920.016, respectively, applying a mass tolerance of 5 ppm.

3. Results and Discussion

3.1. Synthesis and Preparation of d_{70}-Sulfavant A

The synthesis of d_{70}-SULF A was achieved by the ameliorated synthetic strategy of Manzo et al. [10], which was adapted to the preparation of the analog with deuterated fatty acids (Figure 2). In detail, the procedure started with acetylation of D-glucose, followed by selective anomeric deacetylation by benzylamine. Coupling with 1,2-*O*-isopropylidene glycerol by trichloroacetimidate methodology [14,15,19,20] gave 3-*O*-(2′,3′,4′,6′-tetra-acetyl)-β-D-glucosyl-glycerol that was condensed with d_{35}-stearic acid in order to obtain the intermediate 1,2-*O*-d_{70}-distearoyl-3-*O*-β-D-glucosyl glycerol. The sulfonation of the 6′-carbon through an iodinate derivative in agreement with Manzo et al. [10] was performed to finally obtain d_{70}-SULF A after the last hydrazinolysis step.

3.2. Method Development

Many of the studies of ultra- and ultra-high-performance liquid chromatography–mass spectrometry (UPLC- and UHPLC-MS) methods have addressed analysis of lipids [17,21]. However, no specific method for sulfoglycolipids has been reported so far.

UHPLC Kinetex® PS C18 column is a reversed-phase product that is recommended by the manufacturer for the analysis of polar compounds and weakly acidic compounds. We tested several eluting conditions with gradients of methanol with aqueous buffers at different pH and temperatures to optimize the chromatographic result for SULF A and d_{70}-SULF A. The best result in terms of sensitivity, shape of the peak and retention time was obtained with a short gradient (run time 10 min.) of increasing amount of methanol in 0.32 mM ammonium hydroxide solution (0.005%), adjusted to pH 8.0 by acetic acid, at 28 °C. Deuterated analogue of SULF A (**2**), synthesized in-house, was also used during the tests as internal standard (IS). Figure 3 shows chromatograms obtained by base peak extraction of SULF A (analyte) and d_{70}-SULF A (IS). The IS and the analyte eluted with a slightly different retention time due to the isotope effect at 2.9 and 3.0 min., respectively. The quantitation (LOQ) and detection (LOD) limits were 6.5 ngmL^{-1} and 1.9 ngmL^{-1}, respectively, both established in methanol.

3.3. Linearity, Accuracy, Precision and Matrix Effect

The liquid chromatography-mass spectrometry (LC-MS) method was validated for linearity and by an evaluation of intra-and inter-day precision and accuracy in accordance with the currently-approved EMEA and FDA guidelines for the validation of bioanalytical methods [22,23]. Moreover, with the aim to demonstrate the consistency and robustness for the analysis of biological samples, we also tested the above-mentioned parameters in five different replicates of lipid extracts derived from mice lung tissues.

Samples of SULF A and d_{70}-SULF A were tested in triplicates between 1 and 2000 ngmL^{-1}. The resulting calibration curve was found to be linear in the concentration range between

2.5 and 2000 ngmL^{-1}, as shown in Figure 4. Analogously, SULF A and d_{70}-SULF A were spiked at the maximum concentration into MeOH and in a lipid extract of mice lung in MeOH in order to evaluate matrix effect (one of the five extracts available). Samples of the two calibration curves (in pure MeOH and in matrix) were carried out by dilution from the more concentrated solution. Table 2 outlines the slope values and R^2 parameters of the calibration curves acquired at different days (T0 and T48). The R^2 was found to be 0.999 and stable during the days (slight differences in the slope could be appointed to the variability linked to the instrument response). The difference in the mean ordinate between SULF A and d_{70}-SULF A is in agreement with the isotope effect that can cause a different degree of ion suppression. However, although the ionization and chromatographic response was not identical, the peak area ratio of the analyte versus the IS was constant over the range of tested concentrations, with a correlation coefficient (r^2) that is always higher than 0.9991.

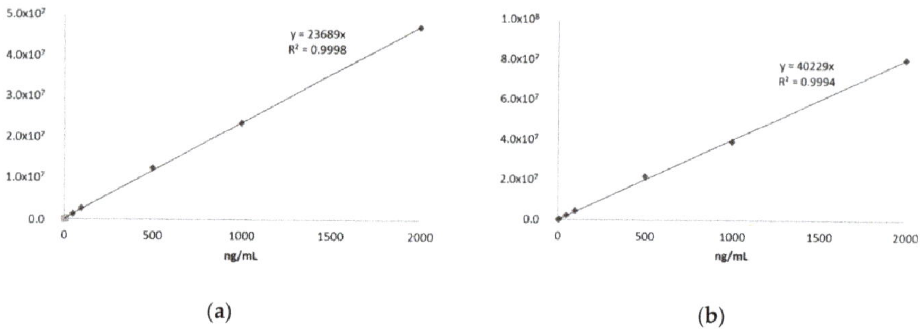

Figure 4. Calibration curve of d_{70}-Sulfavant A (**a**) and Sulfavant A (**b**).

Table 2. Calibration curve of d_{70}-Sulfavant A and Sulfavant A at T0 and T48, both in MeOH and in matrix.

	T0		T48	
	MeOH	**Matrix**	**MeOH**	**Matrix**
SULF A	y = 40229x R^2 = 0.9996	y = 47981x R^2 = 0.9997	y = 50537x R^2 = 0.9991	y = 46508x R^2 = 0.9993
d_{70}-SULF A	y = 23689x R^2 = 0.9998	y = 24546x R^2 = 0.9991	y = 28250x R^2 = 1	y = 24185x R^2 = 0.9996

The matrix effect (ME) is one of the most important parameters to investigate during the development of an analytical method. In order to evaluate the ME, five mice lung lipid extracts were aliquoted and spiked with three different quantities (30, 300, and 1500 ng) of SULF A (Figure 5), while the same known amount of d_{70}-SULF A (50 ng per sample) as IS was added to compare the area of the analyte in matrix with the area measured in pure MeOH (normalized matrix effect-nME). A higher CV was observed for the high concentrations in comparisons with the low ones, as can be seen in Table 3. However, all of the datasets (intra- and inter-day) passed the 15% acceptance criterion, with an overall nME average of 10%. This finding was the first evidence of the robustness of the proposed analytical procedure.

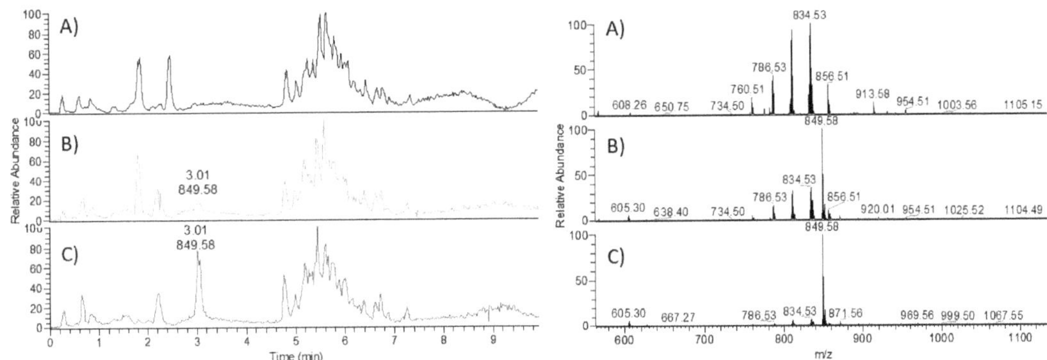

Figure 5. Chromatograms and mass spectrum acquired at 3 min. of pure matrix extract (**A**), matrix spiked with 300 ng (**B**) and 1500 ng (**C**) of Sulfavant A. Chromatograms were obtained as m/z mass range between 400 and 1200.

Table 3. Accuracy and precision of quality control (QC) samples of Sulfavant A during intra-day and inter-day analysis.

	SULF A in MeOH			
	Nominal Concentration (ngmL^{-1})	**Mean ± SD** [a]	**Accuracy (%)** [b]	**Precision (CV %)** [c]
T0	30	27.8 ± 2.2	92.6	8.0
	300	315.0 ± 19.4	105	6.1
	1500	1247.8 ± 69.2	84.2	5.5
T12 (Intra-day)	30	27.7 ± 1.14	92.4	4.1
	300	313.9 ± 17.6	104.6	5.6
	1500	1265.2 ± 27.6	84.3	2.2
T48 (Inter-day)	30	30.7 ± 1.5	102.2	4.9
	300	327.7 ± 16.3	109.2	5.0
	1500	1353.6 ± 15.0	90.2	1.1
	SULF A in Matrix			
	Nominal Concentration (ngmL^{-1})	**Mean ± SD** [a]	**Accuracy (%)** [b]	**Precision (CV %)** [c]
T0	30	26.0 ± 2.2	86.6	8.6
	300	283.1 ± 6.5	94.4	2.3
	1500	1459.9 ± 79.7	97.3	5.4
T12 (Intra-day)	30	28.6 ± 2.0	95.2	7.1
	300	274.4 ± 16.6	91.5	6.1
	1500	1457.3 ± 84.1	97.1	5.8
T48 (Inter-day)	30	27.0 ± 1.2	90.0	4.3
	300	279.8 ± 22.0	93.3	7.9
	1500	1432.5 ± 46.2	95.5	10.2

[a] SD = Standard Deviation [b] [Measured mean concentration/ nominal concentration] × 100. [c] [Standard deviation/ measured mean concentration] × 100.

This analytical method allowed for the detection of SULF A at nanomolar concentration (LOD was below 2.5 ngmL^{-1}) with a LOQ of 6.5 ngmL^{-1}, in both only MeOH and in the presence of matrix. Indeed, at the LOD the analyte in the samples can be identified by only HR-MS, but it cannot be quantitatively determined with appropriate precision and accuracy, whereas, at starting from the calibration point at 5 ngmL^{-1}, the peak of SULF A is also identifiable with MSMS and the signal appears to be defined and reproducible.

Moreover, no carryover effect was detected during analysis of a blank sample after the injection of a standard sample at the highest calibration point (2000 ngmL^{-1}).

The reliability of this UHPLC-MS method was evaluated by measuring intra-day and inter-day accuracy and precision (n = 3 in MeOH; n = 5 in matrix) of SULF A in quality control (QC) samples at three different concentrations (30, 300, and 1500 ngmL^{-1}). Table 3 reports the results in intra-day and inter-day accuracy and precision. The accuracy of QC samples ranged from 84.2 to 109.2 %, and precision ranged from 1.1 to 10.2% (Table 3). The overall assay accuracy (mean of accuracy estimates of all quality control samples) was 96% in MeOH and 93.4% in matrix, and overall assay precision (mean of precision estimates of all quality control samples) was 4.7% in MeOH and 6.4% in matrix.

Revalidation using two different batches of synthetic SULF A (namely, BS90A and BS90A1) was carried out at the same concentrations in the range between 2.5 and 2000 ngmL^{-1} to further support the reliability of the method. The results showed an absolute reproducibility of the analysis with two superimposable calibration curves (BS90A: y = 31579x, R^2 = 0.9999; BS90A1: y = 31331x, R^2 = 0.9999).

4. Conclusions

An UHPLC-MS analytical procedure of SULF A, a 1,2-*O*-distearoyl-3-*O*-β-D-sulfoquinovosylglycerol with promising therapeutic application as a molecular adjuvant, was developed using a new deuterated analog (d_{70}-SULF A) as an internal standard. The method was realized on a high-resolution benchtop Q-Exactive spectrometer using a UHPLC Kinetex® 2.6 µm PS C18 100 Å, LC Column and a gradient elution with water buffered at pH 8 and methanol. The sulfolipids (analyte and internal standard) were quantified in negative ion mode and the calibration curve covers a concentration range of 2.5–2000 ng mL^{-1} with an LLOQ of 6.5 ng mL^{-1}. The new method was validated measuring the precision and accuracy on the quality control samples in the intra-day and inter-day study and its robustness was proven by applying it on a sample matrix.

The efficacy of this analytical method is of general interest to study sulfoglycolipids that are chemically correlated to SULF-A, and suggests that it could be extended for the study of both the physiological and biological role of this class of compounds. Furthermore, the short time of the UHPLC chromatographic analysis and sensitivity makes the procedure adapt to analysis of large number of biological samples.

Pharmacokinetic and drug metabolism tests are mandatory for expediting the progress of compounds with promising properties from discovery to development phase. Mass spectrometry is one of the key technologies in bioanalysis during preclinical and clinical studies [24]. The presented LC-MS method has been designed to provide a mechanistic understanding of the pharmacokinetics, pharmacodynamics, and toxicity of SULF-A. The excellent results in accuracy, sensitivity and reproducibility indicate that the analytical protocol can be utilized in pharmacological and translational research on this negatively-charged lipid, providing accurate results with a high value for further decisions in the drug development.

Author Contributions: Conceptualization, G.N., E.M., G.d. and A.F.; methodology, G.N., A.M.C. and M.Z.; standard preparation, M.Z., L.F. and C.G.; data curation, writing-original draft preparation, G.N.; writing-review and editing, E.M. and A.F. All authors have read and agreed to the published version of the manuscript.

Funding: This work was supported by the project "Antitumor Drugs and Vaccines from the Sea (ADViSE)" project (CUP B43D18000240007–SURF 17061BP000000011) funded by POR Campania FESR 2014-2020 "Technology Platform for Therapeutic Strategies against Cancer"-Action 1.1.2 and 1.2.2.

Acknowledgments: G.N., E.M. and A.F. thank BioSEArch SRL for the supply of sulfavant.

Conflicts of Interest: The authors declare no conflict of interest.

References

1. Nuzzo, G.; Gallo, C.; D'Ippolito, G.; Manzo, E.; Ruocco, N.; Russo, E.; Carotenuto, Y.; Costantini, M.; Zupo, V.; Sardo, A.; et al. UPLC–MS/MS Identification of Sterol Sulfates in Marine Diatoms. *Mar. Drugs* **2018**, *17*, 10. [CrossRef]
2. Wood, P.L.; Siljander, H.; Knip, M. Lipidomics of human umbilical cord serum: Identification of unique sterol sulfates. *Future Sci. OA* **2017**, *3*, FSO193. [CrossRef]
3. Dias, I.H.; Ferreira, R.; Gruber, F.; Vitorino, R.; Rivas-Urbina, A.; Sanchez-Quesada, J.L.; Silva, J.V.; Fardilha, M.; De Freitas, V.; Reis, A. Sulfate-based lipids: Analysis of healthy human fluids and cell extracts. *Chem. Phys. Lipids* **2019**, *221*, 53–64. [CrossRef] [PubMed]
4. Gallo, C.; Nuzzo, G.; D'Ippolito, G.; Manzo, E.; Sardo, A.; Fontana, A. Sterol Sulfates and Sulfotransferases in Marine Diatoms. In *Computer Methods, Part C*; Elsevier BV: Amsterdam, The Netherlands, 2018; Volume 605, pp. 101–138.
5. Mirzaian, M.; Kramer, G.; Poorthuis, B.J.H.M. Quantification of sulfatides and lysosulfatides in tissues and body fluids by liquid chromatography-tandem mass spectrometry. *J. Lipid Res.* **2015**, *56*, 936–943. [CrossRef] [PubMed]
6. Moyano, A.L.; Li, G.; Lopez-Rosas, A.; Månsson, J.-E.; Van Breemen, R.B.; Givogri, M.I. Distribution of C16:0, C18:0, C24:1, and C24:0 sulfatides in central nervous system lipid rafts by quantitative ultra-high-pressure liquid chromatography tandem mass spectrometry. *Anal. Biochem.* **2014**, *467*, 31–39. [CrossRef] [PubMed]
7. Pintado-Sierra, M.; García-Álvarez, I.; Bribián, A.; Medina-Rodríguez, E.; Lebrón-Aguilar, R.; Garrido, L.; De Castro, F.; Fernández-Mayoralas, A.; Quintanilla-López, J.E. A comprehensive profiling of sulfatides in myelin from mouse brain using liquid chromatography coupled to high-resolution accurate tandem mass spectrometry. *Anal. Chim. Acta* **2017**, *951*, 89–98. [CrossRef]
8. Kongmanas, K.; Xu, H.; Yaghoubian, A.; Franchini, L.; Panza, L.; Ronchetti, F.; Faull, K.; Tanphaichitr, N. Quantification of seminolipid by LC-ESI-MS/MS-multiple reaction monitoring: Compensatory levels in Cgt mice. *J. Lipid Res.* **2010**, *51*, 3548–3558. [CrossRef]
9. Manzo, E.; Cutignano, A.; Pagano, D.; Gallo, C.; Barra, G.; Nuzzo, G.; Sansone, C.; Ianora, A.; Urbanek, K.; Fenoglio, D.; et al. A new marine-derived sulfoglycolipid triggers dendritic cell activation and immune adjuvant response. *Sci. Rep.* **2017**, *7*, 1–10. [CrossRef]
10. Manzo, E.; Gallo, C.; Fioretto, L.; Nuzzo, G.; Barra, G.; Pagano, D.; Krauss, I.R.; Paduano, L.; Ziaco, M.; DellaGreca, M.; et al. Diasteroselective Colloidal Self-Assembly Affects the Immunological Response of the Molecular Adjuvant Sulfavant. *ACS Omega* **2019**, *4*, 7807–7814. [CrossRef]
11. Manzo, E.; Fioretto, L.; Pagano, D.; Nuzzo, G.; Gallo, C.; De Palma, R.; Fontana, A. Chemical Synthesis of Marine-Derived Sulfoglycolipids, a New Class of Molecular Adjuvants. *Mar. Drugs* **2017**, *15*, 288. [CrossRef]
12. Ziaco, M.; Fioretto, L.; Nuzzo, G.; Fontana, A.; Manzo, E. Short Gram-Scale Synthesis of Sulfavant A. *Org. Process. Res. Dev.* **2020**, *24*, 2728–2733. [CrossRef]
13. Manzo, E.; Fioretto, L.; Gallo, C.; Ziaco, M.; Nuzzo, G.; D'Ippolito, G.; Borzacchiello, A.; Fabozzi, A.; De Palma, R.; Fontana, A. Preparation, Supramolecular Aggregation and Immunological Activity of the Bona Fide Vaccine Adjuvant Sulfavant S. *Mar. Drugs* **2020**, *18*, 451. [CrossRef]
14. Manzo, E.; Ciavatta, M.L.; Pagano, D.; Fontana, A. An efficient and versatile chemical synthesis of bioactive glyco-glycerolipids. *Tetrahedron Lett.* **2012**, *53*, 879–881. [CrossRef]
15. Manzo, E.; Ciavatta, M.L.; Pagano, D.; Fontana, A. Synthetic strategy for the preparation of bioactive galactoglycerolipids. *Chem. J. Mold.* **2011**, *6*, 27–29.
16. Vial, J.; Jardy, A. Experimental Comparison of the Different Approaches to Estimate LOD and LOQ of an HPLC Method. *Anal. Chem.* **1999**, *71*, 2672–2677. [CrossRef]
17. Cutignano, A.; Luongo, E.; Nuzzo, G.; Pagano, D.; Manzo, E.; Sardo, A.; Fontana, A. Profiling of complex lipids in marine microalgae by UHPLC/tandem mass spectrometry. *Algal Res.* **2016**, *17*, 348–358. [CrossRef]
18. De Nicolò, A.; Cantù, M.; D'Avolio, A. Matrix effect management in liquid chromatography mass spectrometry: The internal standard normalized matrix effect. *Bioanalysis* **2017**, *9*, 1093–1105. [CrossRef]
19. Schmidt, R.R.; Michel, J. Facile Synthesis ofα- andβ-O-Glycosyl Imidates; Preparation of Glycosides and Disaccharides. *Angew. Chem. Int. Ed.* **1980**, *19*, 731–732. [CrossRef]
20. Schmidt, R.R. New Methods for the Synthesis of Glycosides and Oligosaccharides?Are There Alternatives to the Koenigs-Knorr Method? [New Synthetic Methods (56)]. *Angew. Chem. Int. Ed.* **1986**, *25*, 212–235. [CrossRef]
21. Zhao, Y.-Y.; Wu, S.-P.; Liu, S.; Zhang, Y.; Lin, R. Ultra-performance liquid chromatography–mass spectrometry as a sensitive and powerful technology in lipidomic applications. *Chem. Interact.* **2014**, *220*, 181–192. [CrossRef]
22. European Medicines Agency. An unacceptable choice. *Prescrire Int.* **2011**, *20*, 278.
23. Analytical Procedures and Methods Validation for Drugs and Biologics Guidance for Industry Analytical Procedures and Methods Validation for Drugs and Biologics Guidance for Industry. Available online: https://www.fda.gov/regulatory-information/search-fda-guidance-documents/analytical-procedures-and-methods-validation-drugs-and-biologics (accessed on 1 June 2020).
24. Avataneo, V.; D'Avolio, A.; Cusato, J.; Cantù, M.; De Nicolò, A. LC-MS application for therapeutic drug monitoring in alternative matrices. *J. Pharm. Biomed. Anal.* **2019**, *166*, 40–51. [CrossRef] [PubMed]

Review

Mini-Review: Potential of Diatom-Derived Silica for Biomedical Applications

Angela Sardo [1,*], Ida Orefice [1], Sergio Balzano [1,2], Lucia Barra [1] and Giovanna Romano [1,*]

1 Stazione Zoologica Anton Dohrn, Villa Comunale, 80121 Naples, Italy; ida.orefice@szn.it (I.O.); sergio.balzano@szn.it (S.B.); lucia.barra@szn.it (L.B.)

2 Department of Marine Microbiology and Biogeochemistry (MMB), Netherland Institute for Sea Research (NIOZ), Landsdiep 4, 1793 AB Texel, The Netherlands

* Correspondence: angela.sardo@szn.it (A.S.); giovanna.romano@szn.it (G.R.)

Abstract: Diatoms are unicellular eukaryotic microalgae widely distributed in aquatic environments, possessing a porous silica cell wall known as frustule. Diatom frustules are considered as a sustainable source for several industrial applications because of their high biocompatibility and the easiness of surface functionalisation, which make frustules suitable for regenerative medicine and as drug carriers. Frustules are made of hydrated silica, and can be extracted and purified both from living and fossil diatoms using acid treatments or high temperatures. Biosilica frustules have proved to be suitable for biomedical applications, but, unfortunately, they are not officially recognised as safe by governmental food and medical agencies yet. In the present review, we highlight the frustule formation process, the most common purification techniques, as well as advantages and bottlenecks related to the employment of diatom-derived silica for medical purposes, suggesting possible solutions for a large-scale biosilica production.

Keywords: biosilica; diatom frustule; sustainable production; drug delivery

Citation: Sardo, A.; Orefice, I.; Balzano, S.; Barra, L.; Romano, G. Mini-Review: Potential of Diatom-Derived Silica for Biomedical Applications. *Appl. Sci.* **2021**, *11*, 4533. https://doi.org/10.3390/app11104533

Academic Editor: Leonel Pereira

Received: 31 March 2021
Accepted: 14 May 2021
Published: 16 May 2021

1. Introduction

Diatoms are an extremely diverse group of algae, comprising more than 100,000 different species [1]. They are able to colonise a large plethora of aquatic environments, and play a significant role on a global scale in the biogeochemical cycles of carbon and silicon in the water column. Two diatom species, *Thalassiosira pseudonana* and *Phaeodactylum tricornutum*, have been employed as model species for studies of gene expression and regulation, since they were the first species for which the whole genome was fully sequenced [2,3]. Subsequently, genomes have been sequenced from a number of diatoms possessing specific metabolic or physiological features, such as oleaginous (*Fistulifera solaris*), psicrophylic (*Fragilariopsis cylindrus*), araphid (*Synedra acus* subsp. *radians*), oceanic (*Thalassiosira oceanica*), biofilm-forming (*Seminavis robusta*), and heterotrophic (*Nitzschia* sp.) species [4–9]. Apart from their ecological role, diatoms are also suitable for several biotechnological applications. They can be cultured in the laboratory under sterile conditions and controlled temperatures, light irradiance and nutrient concentrations in order to achieve faster growth rates and to promote the accumulation of specialty products. Diatoms have been employed during the last decades for the production of metabolites exhibiting different biological activities and used as sources for cosmetic ingredients [10], food or feed supplements [11–13], fertilizers [14], and sorbents or accumulators for the bioremediation of aquatic environments [15,16]. Microalgae other than diatoms, especially freshwater green algae, also exhibit a great potential in one or more of the abovementioned fields of research.

The true distinctive feature that makes diatoms more suitable than other taxa for biotechnological purposes, is the high proportion of amorphous silica within their cell wall. This natural source of silicon has already shown several advantages, such as its high surface area and biocompatibility, and can be employed for various research fields, especially for biomedical applications after in vitro or in vivo treatments [17]. Diatom-derived silica is also available in huge amounts in aquatic benthic environments, as a consequence of the sedimentation of dead diatom cells.

Currently, diatom biosilica is considered as a suitable biomaterial for metal removal from aquatic environments, as a catalyst support, in optical devices, as a microsensor, and other kinds of applications [18,19]. Since its presence on the market as a device for aquatic remediation and as food-grade products is a pledge of its effectiveness in these fields, the present review is mainly focused on evaluating the potential of diatom biosilica for biomedical applications.

Diatom biosilica is actually exploited, indeed, for its potential as a drug carrier [20] and as a scaffold for bone tissue regeneration [21]. Biosilica-based processes can be considered as low-cost and environmentally friendly alternatives to processes based on artificial structures. While the production of synthetic materials requires the implementation of specific protocols, biosilica carries the advantage of triggering natural and sophisticated structure formation. For example, the employment of diatom-derived biosilica for the development of optical sensors may turn out to be, in the future, more attractive than using synthetic crystals, since it allows control and manipulation of light in a cost-effective way [22]. Biotemplated-based silica can be synthesized by rapid environmentally sustainable methods (solvent-free procedures), thus avoiding the use of hazardous chemicals, and allowing a good control of condensation rates [23].

In the last years, the effectiveness of living or fossil diatom-derived silica for biomedical applications (drug loadings, bone tissue regeneration) has been largely investigated by various research groups, and at least four recent reviews clearly summarize the most relevant studies [18,24–26]. In the present review, we pinpoint the major advantages and bottlenecks related to the employment of diatom silica sources. To this aim, we list examples of diatom-based systems that revealed satisfactory results in the laboratory and might be suitable for scale-up in industrial applications, and the few drawbacks that hinder the use of diatom-derived biosilica as a medical device. We critically compared the effectiveness of diatoms silica sources for drug delivery with respect to other biological and non-biological sources. This review also describes the process of frustule formation, the main techniques of silicon purification, and possible solutions to pave the way to silicon production on an industrial scale.

2. Silicon Capture from the Environment, Transport and Storage, Frustule Formation

In contrast with nitrogen and phosphorus metabolisms, silicon uptake in diatoms is linked with aerobic respiration rather than being strictly related to photosynthesis [27]. Silicon is mainly found in aquatic environments as $Si(OH)_4$, and is usually transformed into solid SiO_2 or other siliceous composites in the presence of organic substances, through condensation reactions [28]. It can enter cells by diffusion across their membranes; nevertheless, at very low concentrations of silicic acid in the surrounding environment, the cells activate silicon transporters that facilitate the uptake [29]. The genes coding for silicic acid transporters (SITs) were isolated and characterized from the marine diatom *Cylindrotheca fusiformis* several years ago [30,31] and, recently, in the freshwater species *Synedra ulna* subsp. *danica* [32]. The SITs specifically transport silicic and germanic acids through the lipid bilayer [33]. Studies performed on *C. fusiformis* revealed that ten transmembrane segments allow the passage of silicic acid through the lipid bilayer membrane, and chemical recognition is likely based on amino acids (Figure 1). The extent of frustule silicification depends on the rate of silicon uptake that is driven, in turn, by both the availability of the substrate and the expression levels of the SIT genes [30]. Five distinct clades have been identified for SIT genes in diatoms. The presence of genes from distinct clades within the same diatom species is likely to reflect different responses to changes in silicon availability and environmental conditions [34].

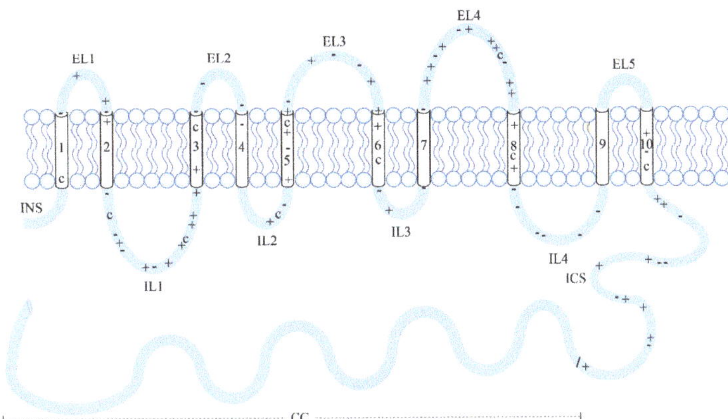

Figure 1. Silicic acid transporters (SITs) in *Cylindrotheca fusiformis*. SITs contain 10 transmembrane segments (white cylinders), which allow the passage of silicic acid through the lipid bilayer membrane; an intracellular amino-terminal segment (INS); and an intracellular carboxy segment (ICS) connected to a coiled-coil motif (CC), which may play a role in the interactions with other proteins. Pluses (+) and minuses (-) indicate the position of positively and negatively charged amino acids, respectively. A major role in silicon uptake is likely played by cysteine residues (C) because of their sulfhydryl blocking agents. Figure redrawn from [31].

Intracellular pools of silicic acid can reach concentrations well beyond the saturation limit (2 mM), and this is likely due to complexation by organic compounds that prevents polymerization. The extent of the internal pools is both species-specific [35] and dependent upon environmental conditions [36]. When silicon deposition and uptake are imbalanced and internal pool tends to increase, a concentration gradient determines the efflux of the "unbound" silicon fraction outside the cell membrane [37].

The diatom cell wall has a Petri dish-like structure, with an upper part, known as epitheca, overlapping the smaller lower part, named hypotheca (Figure 2). Both epitheca and hypotheca consist, in turn, of valves and siliceous girdle bands, which confer an ordered structure to the frustule. Diatom frustule is a composite made of biogenic silica, carbohydrates and glycoproteins [31].

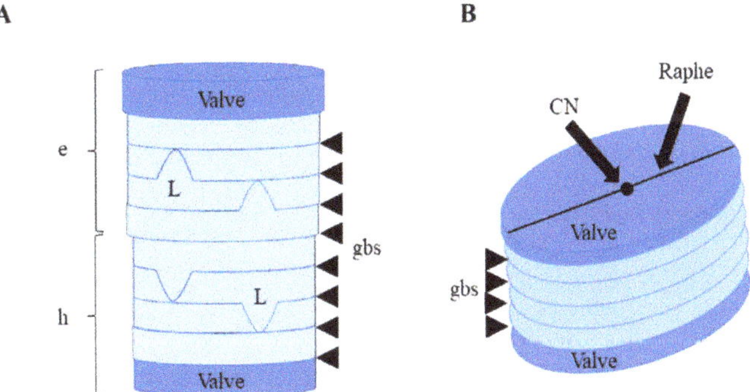

Figure 2. Schematic representation of the cell wall structure of a centric diatom. (**A**) Dark-blue disks represent valves. The upper and larger part is the epytheca (e), the lower and narrower one is the hypotheca (h). Girdle bands (gbs) are indicated by arrows. Ligulas (L) are the bell-shaped structures within girdle bands. (**B**) Diagram of a pennate diatom showing the central nodule (CN) and the raphe on the upper valve. Figure from [38].

Frustule formation occurs through the polymerization of silicic acid in specific compartments, namely, the silica deposition vesicles (SDV), associated to the membrane silicalemma [39,40]. Silicalemma contributes to silicification through both the regulation and recognition of membrane-associated compounds, as well as the formation of a suitable microenvironment for polymerization.

SDVs in diatoms are formed inside the plasma membrane during cell division [41]. During valve formation, the SDVs rapidly expand and their movement is driven by the cytoskeleton [42].

The following several frustule-associated proteins are involved in the formation of the diatom cell wall: frustulins, pleuralins, cingulins, silacidins and silaffins. Aside from proteins, long-chain polyamines, which are constituents of diatom biosilica, are likely involved in silica biogenesis [43].

The following three main levels of cell wall structure organization have been found: (1) microscale, the largest one that determines the outline shape of the valve or girdle band; (2) the mesoscale, at which organized substructures are formed within the SDV; and (3) the nanoscale, which comprises the first products of polymerization and generates different frustule structures/textures of nanometric dimensions ([38] and references therein).

3. Diatom Biosilica Sources

Diatom-derived silica can be obtained either from living cultures or fossil diatoms (diatomite, e.g., chalky deposits of skeletal remains). The energy required for diatom growth is sustained by either led-based (i.e., low energy demanding) artificial light or sunlight. Furthermore, the nutrients required for algal growth, such as nitrates, phosphates, silicates, vitamins, and some trace elements, can be purchased for a relatively cheap price or even obtained from wastewaters. To avoid both the costs of artificial illumination and the seasonal variability of sunlight, cells can also be grown heterotrophically [44–47], although organic substrates are to be supplied in this case. However, only a small number of species are able to grow in the dark [48,49], and organic compounds can promote bacterial growth leading to culture contaminations and to a decrease in cell growth. Biosilica is obtained after cell dewatering (i.e., centrifugation or filtration of the whole culture), followed by a purification process that is usually based on treatments with strong acids and/or high temperatures (see below). Besides, the limited motility of diatoms (due to the lack of flagella) and the "heavy" cell wall (due to the presence of a high silicon amount) enhance the spontaneous sinking of cells, limiting the volume to harvest and, thus, costs of biomass collection.

Diatoms generally exhibit fast growth rates and high lipid and biomass productivities, [50] which can be further enhanced by tuning growth conditions [51,52], making diatoms promising candidates for mass culturing. However, to the best of our knowledge, no diatom-based industrial plants (i.e., indoor or outdoor systems of algal culturing) are focusing on biosilica production as their main activity. Follow-up studies are thus required to lay the foundations for the industrial production of silica-based biomaterials.

The most abundant source of biosilica that does not foresee the induction of living cultures is diatomite, which can be easily crushed into a fine powder to become a marketable product, namely, diatomaceous earth (DE). Diatomite is made of frustules of dead diatom cells, usually found in benthic environments. The harvesting of fossil frustules, which are naturally present in benthic environments, is cost-effective and makes diatomite a promising starter for the industrial production of biosilica. However, the composition of DE is variable and the purity is often lower than that of living culture-derived frustules. The quality and abundance of these impurities vary upon environmental and aging conditions [18]. DE, generally made of ca. 80–90% of silicon and of clay minerals [53], is used as a raw material for different kinds of applications, such as agricultural fertiliser, sorbent for pollutants, and filler in plastics and paints to improve the strength of construction materials. In addition, DE is also employed to filter impurities and as an abrasive agent in cleaning and polishing products.

4. Frustule Cleaning/Purification: Main Techniques and Technical Issues

Frustules can be thus purified from both living culture-derived algal biomass and diatomite stocks. The impurities of diatom frustules mainly consist of organic matters adhered to their surface [54]. In the case of diatomite samples, impurities are present in larger amounts, and can vary in relation to the local environment and aging conditions of these natural stocks [18]. Diatomite impurities typically contain also clay and metallic oxides, such as aluminium and ferric oxides [55]. Before cleaning procedures, diatomite particles usually undergo a first step of pulverization, in which micrometric powder is grinded to nanoparticles by mechanical crushing and sonication. However, apart from a few exceptions, most studies report purification protocols based on raw material derived from living cultures rather than diatomite, which is currently the only diatomic silica-based marketable product.

Organic impurities can be removed from the silica frustule by either a chemical pre-treatment with acids or other oxidative agents, or by exposing the frustules to high temperatures. Some studies, aimed at assessing the efficacy of preliminary hydrochloric acid treatments for organic mass removal, showed that acid concentration greatly influenced both the removal rate of impurities and the state of preservation of the frustule shape, with strong acidic pre-treatments causing frustule erosion [56]. Potassium permanganate can be also used to pre-treat frustules for organic compound removal [57,58]. However, this procedure is essentially limited to remove impurities outside the frustule, and pre-treatments with acidic solutions are usually applied (even if they are not mandatory) when purification protocols do not foresee acid-based cleaning procedures, such as baking-based purifications [59]. Some preliminary oxidations with acid solutions do not exclude the employment of both acids and high temperatures. Treatment of diatom frustules with sodium permanganate and oxalic acid, for example, is followed by perchloric acid treatments at 100 °C [57].

Baking (i.e., strong heating of silica cell walls) of diatom frustules at 400–800 °C is the simplest and least expensive method to remove organic components. However, high-temperature treatments can alter diatom architecture and pore size [60]. Oxygen plasma etching, a procedure consisting of the removal of impurities using ionised gases, was found to be effective to preserve the frustule structure, with a negligible loss of material and without shape alterations [61,62].

The most commonly used procedure for the removal of organic matter and the purification of diatom biosilica is, however, an oxidative washing treatment. Some protocols require the use of 30% [54,63–67] or 15% [68] hydrogen peroxide solutions.

The most common washing solvents used in acid-based treatments of diatom frustules are sulphuric [69,70] and nitric [68,71] acids. Sulphuric acid treatment is rapid (10–30 min) and revealed successful even on small amounts of biosilica [55]. Despite the rapidity of this strong acid-based method, cleaning procedures are time-consuming, since several washes with distilled/deionised water are required for a complete acid removal. However, the effect of acid strength needs to be evaluated in each case, since silica nanostructures can be damaged by the action of acids. For example, frustules from poorly silicified diatom species can be dissolved in strong acid cleaning solutions [70].

To improve the efficiency of biosilica purification, Wang and co-workers [72] set up a vacuum cleaning method in which all the cleaning steps, which are cell extraction, acid treatment and washing, are carried out on polytetrafluoroethylene (PTFE) filter cloths, thus decreasing the processing time. This allows the recycling of the sulphuric acid used for cleaning, decreasing the amount of both the reagent needed for purification and the liquid wastes. The main drawback of the vacuum cleaning method is that it depends on the mechanical properties of the raw material, and cannot be applied on poorly silicified diatoms.

Some purification methods combine the use of both sulphuric acid and hydrogen peroxide in a strong oxidizing agent (2 M H_2SO_4, 10% H_2O_2) called Piranha solution [26,73]. The purification process is relatively fast, while post-treatment washes can be time-consuming.

The removal of Piranha solution requires, indeed, an overnight treatment with HCl (5 M, 80 °C) and two further washes with distilled water to eliminate the HCl residuals [20]. The main treatments for frustule separations, the tested diatom silica sources, and the main bottlenecks of each cleaning technique are summarized in Table 1.

Table 1. Pre-treatments and treatments for diatom frustule cleaning and their main advantages and drawbacks.

	Treatment	Principle for Organic Matter Removal	Diatom Species	Diatom Silica Source	Advantages	Drawbacks	Reference(s)
Pre-treatments	HCl	oxidizing washing	*Nitzschia closterium*, *Thalassiosira* sp.	freeze-dried samples	high purity of frustules	possible frustule erosion depending on acid strength	[56]
	$KMnO_4$ + $C_2H_2O_4$	oxidizing washing	*Fragilariopsis cylindrus*, *Fragilariopsis kerguelensis*, *Pseudonitzschia seriata*, *Thalassiosira nordenskioeldii*, *Thalassiosira aestivalis*, *Thalassiosira pseudonana*, *Thalassiosira weissflogii*	wet pellets washed with sodium lauryl sulfate	no frustule erosion	removal of the only external organic matter	[57,58]
Treatments	baking	high temperature	*Navicula* sp.	APS-fuctionalised diatoms on a mika surface	reduction in hazardous chemicals	possible alterations of pore size, possible post-treatments with acid solutions	[60]
	low-temperature plasma ashing	ionised gas	*Navicula*, *Amphora*, *Cocconeis*, *Planothidium* spp.	desalted drops of cultures, freeze-dried samples	no frustule dissolution	unsuitable for saltwater species, expensive, post-treatments with hazardous chemicals	[61,62]
	H_2O_2	oxidation	DE, *Ni tzschia frustulum*, *Pinnularia* and *Coscinodiscus* spp., *Thalassiosira pseudonana*, *Cylindrotheca closterium*	desalted and freeze-dried cultures, diatom composites	less dangerous than strong acids	long incubation, high-temperature post-treatments needful to increase efficiency	[54,62–67]
	H_2SO_4	strong oxidation	*Thalassiosira rotula*, *Coscinodiscus wailesii*	living cultures	high efficiency in organic matter removal	hazardous chemicals, dissolution of thin frustules, time-consuming post treatments	[69,70]
	H_2SO_4 + PTFE filters	strong oxidation under vacuum	*Nitzschia*, *Ditylum*, *Skeletonema*, *Coscinodiscus*	living cultures on a filter cloth	reduced acid amounts	unsuitable for thin frustules	[72]
	HNO_3	strong oxidation	*Pinnularia* sp., *Coscinodiscus concinnus*	harvested cells	high efficiency in organic matter removal	high-temperature treatments needful to increase efficiency	[68,71]
	Piranha solution (H_2SO_4 + H_2O_2)	strong oxidation	*Thalassiosira pseudonana*	PBS-washed cells	high efficiency in organic matter removal	time-consuming post-treatments	[73]

5. Silica for Biomedical Applications: Advantages

The main benefits of biosilica for biomedical purposes are as follows: plasticity of frustules for functionalization, biocompatibility, possibility of genetic transformation of living cultures for protein immobilization, and high availability of silica-derived diatoms.

The biosilica derived from diatoms requires cheap synthesis processes [26], and is also characterised by chemical inertness, low or null toxicity, thermal stability and high availability [18]. Silica has been widely investigated in drug delivery systems because of its high robustness and versatility compared to other materials [74], and frustules derived from both living cultures and diatomite particles have successfully been employed as drug carriers [73,75].

5.1. Surface Functionalization for Drug Loading and for Biosensing Chips for Biomedical Applications

Frustule functionalization consists of modifying its surface to enable the formation of stable covalent bonds with proteins or DNA [26,76], by introducing chemically reactive species functioning as cross-linkers. This step is crucial to improve the quality of the resulting material for specific applications. Chemical modification of biosilica can be critical, for example, to regulate the kinetics of drug release, and the high surface-to-volume ratio makes this raw material particularly suitable for drug delivery. Diatom frustules are characterized by precise and species-specific cell morphologies, and both the size and shape can highly differ among distinct diatom taxa. It has been estimated that the surface area ranges between 1.4 and 51 m^2 g^{-1} [77–80]. The size and the architecture of the pores are likely to influence drug release [75].

Drug release in biosilica-based systems is usually characterized by the following two phases: a first phase of fast release, due to the detachment of drug molecules weakly bound to the frustule surface, and a slow releasing phase, due to drug delivery from the internal pore structure of diatom frustules [81]. Chemical modifications of diatom-derived biosilica allow their use as a carrier of both soluble and insoluble drugs.

The effectiveness of DEs as delivery systems for the drugs gentamicin (soluble) and indomethacin (insoluble) was demonstrated in previous studies [67], in which DE was modified with a self-assembling monolayer (SAM) including organosilanes and phosphonic acids, thus rendering the diatom frustules hydrophilic or hydrophobic, respectively, before drug loading. A sustained release of indomethacin, which has been exploited as a model drug for silica-based devices, was also demonstrated with DE particles functionalised by dopamine-modified iron oxide nanoparticles (DOPA/Fe_3O_4 nanoparticles). Diatom-derived silica was employed, in this case, as a magnetically guided micro-carrier for drug delivery, since dopamine amino groups on the diatom surface allow the attachment of targeting biomolecules [78]. Another kind of functionalization can be obtained by combining the frustule with graphene oxide (GO) sheets through covalent bindings. These nano-hybrid composites are suitable drug microcarriers. GO sheets enhanced, indeed, drug-surface interactions, improving the kinetics of drug release [82].

Silica functionalization was also used to counteract cancer progression, through the delivery of water-insoluble antitumor drugs. A recent study showed that DE particles coated with vitamin B12 allowed better delivery of cisplatin and 5-fluorouracil (5-FU), two anticancer agents effective against colorectal cancer cells [83]. Silicon nanoparticles (SiNPs) were also functionalized with 5-FU and the chemopreventive agent curcumin, and then encapsulated into acid-resistant microspheres to show the effectiveness of oral administration of these chemotherapeutics against colorectal cancer [84].

DE particles were also used as a solid drug-carrier in phospholipid suspensions for new oral formulations of non-anticancer water-insoluble drugs, such as the anticonvulsive carbamazepine [85].

While the abovementioned applications of biosilica were all based on the employment of fossil sources, other studies were focused on culture-derived biosilica. Functionalised frustules of the diatom *Nitzschia palea* have been successfully exploited as carriers for the antibacterial complex tyrosine-Zn(II); zinc ions covalently bounded to the frustule surface showed, indeed, a toxic effect on bacteria, thus reducing their concentration [86]. Esfandyari et al. [87] exploited the potential of *Chaetoceros* sp. frustules to detect circulating tumour cells. Diatoms were magnetized with iron oxide nanoparticles, and then conjugated with the monoclonal antibody Trastuzumab; this system was effective in selectively targeting and separating breast cancer cells, SKBR3 cells (HER2 positive cells), from HER2-negative cells under a magnetic field. The optical properties of these diatoms allowed to detect this specific binding ability by fluorescence microscopy, thanks to the optical properties of the silica.

Similar studies on antibody-functionalized nanoparticles deriving from living cultures were already performed more than ten years ago, and they exploited the potential of two modified centric diatoms as photoluminescent biosensors. Functionalization of *Coscinodiscus wailesii* frustules was one of the pioneer studies highlighting antigen recognition from antibodies that had been covalently bound to frustules [88]. Gale and co-workers [89] succeeded in transforming *Cyclotella* sp. frustules with the model rabbit IgG antibody, showing a correlation between the photoluminescence associated with the frustule/antibody complex and the antigen (goat anti-rabbit IgG) concentration. The main types of diatom silica functionalization are summarised in Table 2.

Table 2. Sources, type of functionalization and biomedical applications of diatom-derived biosilica.

Diatom Source	Type of Functionalization	Main Application	Aim	Reference(s)
Coscinodiscus wailesii	Silanization and antibody conjugation	Biosensor	Specific recognition antigen–antibody (murine monoclonal antibody)	[88]
Coscinodiscus wailesii	Silanization and antibody conjugation	Biosensor	Tethering and detecting antibodies (mix of normal rabbit serum and purified Ig-Y)	[64]
Cyclotella sp.	Silanization and antibody conjugation	Biosensor	Selective and label-free photoluminescence-based detection of antigen–antibody (IgG-rabbit) complex formation	[89]
Chaetoceros sp.	Iron oxide nanoparticles and antibody conjugation	Biosensor (with magnetic properties)	Selective targeting of SKBR3 cancer cells through the employment of antibody (Trastuzumab) bioconjugation	[87]
Thalassiosira weissflogii	Nitroxide 2,6,6-tetramethylpiperidine-N-oxyl (TEMPO) conjugation	Drug carrier	Ciprofloxacin delivery in fibroblasts and osteoblasts	[90]
Aulacoseira sp.	Silanization, and oligo (ethylene glycol) methacrylate copolymers addition	Drug carrier	Improvement of levofloxacin delivery	[75]
Nitzschia palea	Amino acid (Tyr-Zn^{II}) conjugation	Drug carrier	Inhibition of bacterial growth	[86]
Diatomaceous earth	Silanization and phosphonic acids conjugation—self-assembling monolayer	Drug carrier	Improvement of indomethacin and gentamicin delivery	[67]
Diatomaceous earth	Silanization and phosphonic acids modifications	Drug carrier	Improvement of indomethacin delivery	[91]
DE mineral rocks	Graphene oxide, silanization	Drug carrier	Improvement of indomethacin delivery	[82]
Diatomaceous earth	Dopamine modified iron-oxide nanoparticles (DOPA/Fe_3O_4)	Drug carrier (with magnetic properties)	Improvement of indomethacin delivery	[78]
Diatomaceous earth	vitamin B12 and ruthenium (II) complex	Drug carrier	Improvement of the anticancer tris-tetraethyl [2,2′-bipyridine]-4,4′-diamine–ruthenium (II) complex delivery (tested on HT-29 and MCF-7 cancer cells)	[83]
Calcined diatomite	Silanization and siRNA conjugation	Drug carrier	Vehiculating siRNA into tumour cells to downregulate the expression of cancer-associated genes (tested on murine A20 lymphoma cells)	[92]
Calcined diatomite	Silanization and siRNA conjugation	Drug carrier	Vehiculating siRNA into tumour cells to downregulate the expression of cancer-associated genes (tested on H1355 cancer cells)	[93]

5.2. Biocompatibility

Diatom-derived biosilica has several advantages compared to other porous materials, in terms of high compatibility with biological systems [18,26]. Biocompatibility tests were performed on various tumour cells, and some significant examples are reported below. An ATP-based luminescent assay aimed at detecting the short-time (6–24 h) detrimental effects on cells showed that DE particles had very low toxicity on the following three colon cancer cell lines: Caco-2, HT-29, and HCT-116 [94]. The effect of amino-modified DE nanoparticles on human lung epidermoid carcinoma cells (H1355) was evaluated by the MTT (3-(4,5-dimethythiazol-2-yl)-2,5-diphenyl tetrazolium bromide) assay. Different concentrations of diatom particles were tested for 24, 48 and 72 h, and the results showed very low cytotoxicity against the abovementioned tumour cells. This feature made functionalized DE particles useful carriers to transport small interfering ribonucleic acid (SiRNA) inside human lung epidermoid carcinoma cells (H1355), silencing gene expression [93]. Biocompatibility was also assessed on bone cells, such as normal human dermal fibroblasts (NDHS) and Saos-2 osteoblasts, by the functionalization of *Thalassiosira weissflogii* frustules with 3-mercaptopropyl-trimethoxysilane (MPTMS). The mercapto-coated biosilica successfully stimulated the growth of both cell lines, even more than bare cells [90].

The biological compatibility of silica-derived diatoms was also assessed in studies aimed at targeting the antiapoptotic factor B-cell lymphoma/leukemia 2 (Bcl2) with small interfering RNA (siRNA). Specifically, the amino groups of silanized silica particles were complexed with siRNA to downregulate the expression of tumour-associated genes. The target line was the A20 murine lymphoma, and no differences in cytotoxicity between the functionalised frustules and controls (e.g., untreated cells) were observed by applying the following three different methodologies: MTT, Cell-Titer GLO and propidium iodide assays [92].

Biocompatibility between functionalized DE particles and breast cancer cells (lines MCF-7 and MDA-MB-231) has also been proven. In this case, amino-modified particles were further improved by PEGylation (i.e., diatom-coating with polyethylene glycol) and cell-penetrating peptide (CPP) bioconjugation, to promote cell internalization through physical and biological changes in the silicon source. The biological compatibility was also evaluated with a luminescent cell viability assay based on the adenosine triphosphate concentration, and the results showed that the cytotoxicity of biosilica that underwent a double modification with PEG and CCP was lower than that of the bare material, as well as that of diatoms that had been amino-modified only [95].

Most cytotoxicity assays mentioned above were performed on short timescales. The effect of longer exposure times (21 days) was assessed on human embryonic kidney cells (HEK-293) and MDA-MB-231 breast cancer cells exposed to syntherized (e.g., fused at high temperature) diatoms. Biocompatibility was tested through viability assays with the dye Calcein-AM (its fluorescence intensity depends on the activity of cellular esterases, and thus of viable cells), and the results confirmed that natural silicon is not toxic. This suggested that fused diatom frustules could be a suitable alternative for synthetic bone graft substitutes [96]. In order to foresee the effects of long-term exposure of silica-based devices on biological systems, Terracciano and co-workers [97] investigated the in vivo impact of diatomite particles on the model organism *Hydra vulgaris*. Untreated specimens and animals exposed to bare frustules and to diatom nanoparticles modified with the cell-penetrating peptide [(aminooxy)acetyl]-Lys-(Arg)9 (to enhance cellular uptake) were monitored for 14 days, and no detrimental effects in terms of growth rates and apoptosis were observed in all conditions.

In our opinion, further studies on living organisms are mandatory to definitely ascertain the lack of toxicity of biosilica, especially in the perspective of concrete biomedical applications for drug loading and as scaffolds for bone regeneration.

5.3. Employment of Genetically Engineered Diatom Frustules for Protein Immobilization

Diatom particles can be considered as useful scaffolds for enzyme immobilization that could enhance protein properties. Genetic engineering represents a viable alternative to in vitro immobilization systems, as it does not require protein purification and is carried out under physiological conditions [74]. Since silaffins and cingulins are involved in silica condensation becoming part of diatom frustules, the fusion of an exogenous protein to these frustule-associated proteins can result in the strong binding of exogenous proteins to the silica cell wall.

Transformation of diatom genomes with recombinant genes is a useful tool to allow the fusion between enzymes and cell wall proteins. This technology is mentioned in a recent study as living diatom silica immobilization (LiDSI), and has been mostly performed on the model species *T. pseudonana* [98,99]. To our knowledge, the pioneer studies focused on enzymes immobilised on diatom biosilica were aimed at inserting and blocking the bacterial enzyme hydroxylaminobenzene mutase (HabB) on the silaffin tpSil3 of *T. pseudonana* frustule [98]. Aside from the potential of this specific genome modification, this study paved the way for the genetic manipulation of diatom species to enhance protein immobilization on frustules for biomedical purposes.

The genome of *T. pseudonana* has been recently modified with the insertion of exogenous genes encoding the fusion of two enzymes, glucose oxidase and horseradish peroxidase, with cell wall proteins, enabling a regioselective functionalization, and suggesting that silica morphology could influence the effectiveness of the enzymes reactivity [99]. The frustule of this species has been also antibody-functionalised, in order to test its effectiveness in binding large and small antigen molecules [100].

5.4. Availability of Biosilica Feedstocks

In contrast with other synthetic materials, diatom biosilica is already available in huge amounts as diatomite. Moreover, diatom-derived silica feedstock could be easily obtained by culturing these microalgae in open ponds or enclosed systems, and separating them from the organic matter after culture dewatering.

6. Silica for Biomedical Applications: Bottlenecks

Among the weak aspects of the production of biosilica-based devices, it is worth mentioning that diatom frustules require strong treatments that are usually based on toxic and/or dangerous chemicals [26,70,73,101]. Furthermore, the accuracy of the purification protocols should to be the highest possible in case the biosilica-based material is to be used for biomedical purposes. Besides, experiments based on living culture manipulations are often carried out under axenic conditions [87,89], which requires very careful maintenance to keep the strains bacteria-free and, in case of contamination, highly meticulous protocols to remove bacteria need also to be applied [102].

Another drawback of using biosilica for biomedical purposes is the low degradation rate of this material. Although biosilica is known to be less stable than crystalline silica and its dissolution rate can be further enhanced by specific physico-chemical manipulations [103], it can still persist for long periods within organs with limited blood supply [104], which may lead to detrimental health consequences. Some authors do not consider biosilica degradation as a real problem, since the silicic acid (i.e., the main product of silica degradation) is naturally found in human tissues and can be easily excreted from the kidneys [105,106].

However, the employment of diatom-based biosilica for biomedical purposes has not been approved yet by the Food and Drug Administration (FDA), and neither by other safety governmental agencies. The approval of diatom-based biosilica for biomedical purposes is mandatory to follow up all the studies which foresee the development of biosilica-based devices for biomedical purposes. The FDA considers amorphous silica less toxic than crystalline forms and silicates as generally recognized as safe (GRAS) materials [57,107], and established that they can be included in oral delivery ingredients in amounts up to 1500 mg per day [103]. Some diatom-derived sources of silicon can be treated, indeed, to reach the food grade and can be used for nutraceutical purposes. Currently, diatomite of certain "purity levels" is considered as a food additive and is permitted as animal feed (https://www.accessdata.fda.gov/scripts/cdrh/cfdocs/cfcfr/CFRSearch.cfm?fr=573.340, accessed date: 22 March 2021), as inert carrier or anticaking agent, and as additive for pharmaceutical preparations. Unfortunately, the use of diatomite for biomedical applications has not been approved yet.

The recognition of diatomite as a suitable biomaterial for medical and pharmaceutical purposes by the FDA is, indeed, mandatory to pave the way to industrial applications of diatomite-based materials, which have proven to be successful for drug delivery [74], through genetic engineering [108,109] and for regenerative medicine [26].

7. Future Perspectives

In the present review, we report studies from different research groups aimed at assessing the effectiveness of diatom-derived silica for biomedical applications. Previous studies clearly demonstrated that diatom biosilica could be suitable as a drug carrier, as a scaffold for regenerative medicine, and for in vivo enzyme immobilization. We summarized the main benefits and disadvantages associated with the use of this biomaterial, and, in our opinion, the advantages outweigh the drawbacks. Future research should ascertain long-term compatibility of biosilica with biological systems, and should search for cost-effective and environmentally friendly techniques for biosilica production.

The employment of silica-based diatom cell walls for biotechnological applications implies that naturally abundant fossil sources (e.g., diatomite) can be considered as a viable and cost-efficient alternative to synthetic silica, which requires time-consuming and hazardous methods for its production. Overall, fossil biosilica sources can be used as heavy metal sorbents, food-grade additives, fertilizers or biosensors.

In contrast, living diatom-derived biosilica appears to be more suitable than biosilica of fossil origin for biomedical applications such as drug carriers. Specifically, drugs need to be released from the carrier at constant and known rates, and for this purpose the size, shape, and porosity of biosilica particles should vary as little as possible. Biosilica derived from monospecific cultures has, indeed, a regular and predictable structure compared to fossil sources that typically encompass a given morphological diversity, leading to heterogeneity in the frustule size, shape and porosity. A recent device for biosilica production from living cultures has been patented [110] and paves the way to the exploitation of diatoms.

With respect to other biological silica sources, such as sponges and Radiolaria, diatoms can be easily cultured in the laboratory, and are ubiquitous and highly abundant in the marine environment. Moreover, silica structures from diatoms generally have smaller dimensions than protozoan shells and sponge spicules, which can reach 2 mm in length [74]. The reduced particle size is, for us, advantageous in drug delivery systems to enhance degradability, and for avoiding or limiting pulverization pre-treatments. In our opinion, treatments for the depigmentation and elimination of collagen (in which spicules are enveloped) can render the process of sponge-derived silica even more difficult than the typical techniques for frustule purification [111,112]. Other non-biological silica sources, such as fly ash derived from power plants, may be used in lieu of fossil diatom-derived sources. Similarly to diatomite, this alternative source is disposable and abundant in the natural environment, and its effectiveness as a raw material for carbon sequestration [113] and as a biofertilizer [114] has already been demonstrated. Nevertheless, in our opinion, the high heterogeneity of fly ash (structural features vary according to the power plant and the combustion processes), and the presence of metals and other impurities makes it more suitable for bioremediation and land fertilization purposes, rather than biomedical applications.

To improve the economic viability of massive diatom culturing for biosilica production, other fractions of algal biomass could also be exploited. Diatoms can be used, indeed, as photosynthetic biorefineries [115] for a number of industrial processes, and the exploitation of both the inorganic and organic fractions of the algal biomass can contribute to minimise waste production. A possible route for the complete exploitation of diatom biomass is shown in Figure 3. The extraction of the biochemical components from microalgal biomass does not affect the structure and the integrity of silica frustules, which can remain unaltered also in the presence of acidic conditions. Massive diatom culturing can lead to the combined production of biosilica for biomedical applications, and highly valuable organic compounds such as PUFAs and carotenoids.

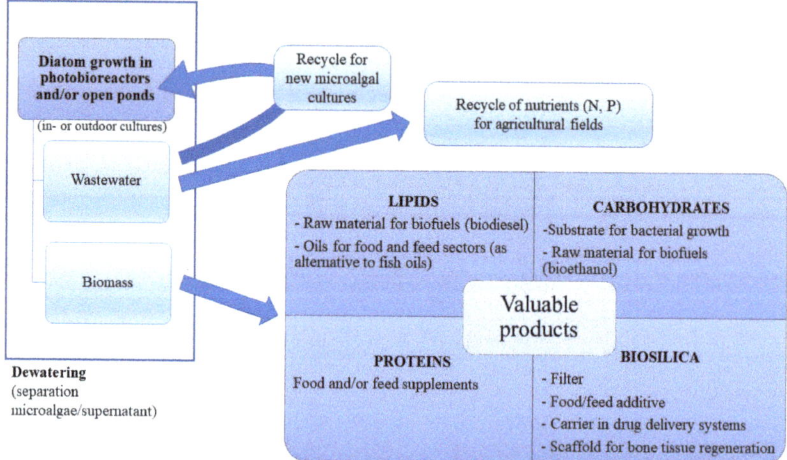

Figure 3. Schematic representation of a hypothetic diatom-based biorefinery for the whole exploitation of microalgal mass. Diatoms are cultured in open ponds or photobioreactors; after harvesting, the remaining water is still rich in nutrients and can be partially reused as medium for new culture inocula or serve as agricultural fertiliser. Four major valuable products, lipids, proteins, carbohydrates, and biosilica can be obtained from the harvested biomass.

In summary, we believe that further studies are required to exclude any acute and long-term toxicity of diatom biosilica—as suggested by Castillo and Vallet-Regi [116]—and this will pave the way for clinical trials of biosilica transplantation. However, assessing the best processes to minimize costs and wastes for the concomitant production of biosilica and highly valuable products is mandatory to lay the foundations for this new industrial application.

Author Contributions: A.S. conceptualization and original draft preparation, I.O. support in draft preparation, figure preparation, support in literature research, S.B. and L.B. editing and critical review, G.R. editing and supervision. All authors have read and agreed to the published version of the manuscript.

Funding: This research received no external funding.

Institutional Review Board Statement: Not applicable.

Informed Consent Statement: The authors did not need any signed informed consent.

Data Availability Statement: Not applicable.

Acknowledgments: The authors are grateful to F. Palumbo and M. Perna (SZN, Marine Biotechnology Department) and G. Lanzotti (SZN, Research Infrastructures for MArine biological Resources, RIMAR) for graphical support. Thanks are due also to Adrianna Ianora for editing and reviewing the English language.

Conflicts of Interest: The authors declare no conflict of interest.

References

1. Mann, D.G.; Vanormelingen, P. An inordinate fondness? The number, distributions, and origins of diatom species. *J. Eukaryot. Microbiol.* **2013**, *60*, 414–420. [CrossRef]
2. Armbrust, E.V.; Berges, J.A.; Bowler, C.; Green, B.R.; Martinez, D.; Putnam, N.H.; Zhou, S.G.; Allen, A.E.; Apt, K.E.; Bechner, M.; et al. The genome of the diatom *Thalassiosira pseudonana*: Ecology, evolution, and metabolism. *Science* **2004**, *306*, 79–86. [CrossRef] [PubMed]
3. Bowler, C.; Allen, A.E.; Badger, J.H.; Grimwood, J.; Jabbari, K.; Kuo, A.; Maheswari, U.; Martens, C.; Maumus, F.; Otillar, R.P.; et al. The *Phaeodactylum genome* reveals the evolutionary history of diatom genomes. *Nature* **2008**, *456*, 239–244. [CrossRef] [PubMed]
4. Lommer, M.; Specht, M.; Roy, A.S.; Kraemer, L.; Andreson, R.; Gutowska, M.A.; Wolf, J.; Bergner, S.V.; Schilhabel, M.B.; Klostermeier, U.C.; et al. Genome and low-iron response of an oceanic diatom adapted to chronic iron limitation. *Genome Biol.* **2012**, *13*. [CrossRef] [PubMed]
5. Galachyants, Y.P.; Zakharova, Y.R.; Petrova, D.P.; Morozov, A.A.; Sidorov, I.A.; Marchenkov, A.M.; Logacheva, M.D.; Markelov, M.L.; Khabudaev, K.V.; Likhoshway, Y.V.; et al. Sequencing of the complete genome of an araphid pennate diatom *Synedra acus* subsp. *radians* from Lake Baikal. *Dokl. Biochem. Biophys.* **2015**, *461*, 84–88. [CrossRef]
6. Tanaka, T.; Maeda, Y.; Veluchamy, A.; Tanaka, M.; Abida, H.; Marechal, E.; Bowler, C.; Muto, M.; Sunaga, Y.; Tanaka, M.; et al. Oil accumulation by the oleaginous diatom *Fistulifera solaris* as revealed by the genome and transcriptome. *Plant Cell* **2015**, *27*, 162–176. [CrossRef]
7. Mock, T.; Otillar, R.P.; Strauss, J.; McMullan, M.; Paajanen, P.; Schmutz, J.; Salamov, A.; Sanges, R.; Toseland, A.; Ward, B.J.; et al. Evolutionary genomics of the cold-adapted diatom *Fragilariopsis cylindrus*. *Nature* **2017**, *541*, 536–540. [CrossRef] [PubMed]
8. Osuna-Cruz, C.M.; Bilcke, G.; Vancaester, E.; De Decker, S.; Bones, A.M.; Winge, P.; Poulsen, N.; Bulankova, P.; Verhelst, B.; Audoor, S.; et al. The *Seminavis robusta* genome provides insights into the evolutionary adaptations of benthic diatoms. *Nat. Commun.* **2020**, *11*, 3320. [CrossRef]
9. Pendergrass, A.; Roberts, W.; Ruck, E.C.; Lewis, J.A.; Alverson, A.J. The genome of a nonphotosynthetic diatom provides insights into the metabolic shift to heterotrophy and constraints on the loss of photosynthesis. *BioRxiv* **2020**. [CrossRef]
10. Mourelle, M.; Gómez, C.; Legido, J. The Potential Use of Marine Microalgae and Cyanobacteria in Cosmetics and Thalassotherapy. *Cosmetics* **2017**, *4*, 46. [CrossRef]
11. Cui, Y.; Thomas-Hall, S.R.; Schenk, P.M. *Phaeodactylum tricornutum* microalgae as a rich source of omega-3 oil: Progress in lipid induction techniques towards industry adoption. *Food Chem.* **2019**, *297*, 124937. [CrossRef]
12. Pudney, A.; Gandini, C.; Economou, C.K.; Smith, R.; Goddard, P.; Napier, J.A.; Spicer, A.; Sayanova, O. Multifunctionalizing the marine diatom *Phaeodactylum tricornutum* for sustainable co-production of omega-3 long chain polyunsaturated fatty acids and recombinant phytase. *Sci. Rep.* **2019**, *9*, 11444. [CrossRef] [PubMed]
13. Tiwari, A.; Melchor-Martinez, E.M.; Saxena, A.; Kapoor, N.; Singh, K.J.; Saldarriaga-Hernandez, S.; Parra-Saldivar, R.; Iqbal, H.M.N. Therapeutic attributes and applied aspects of biological macromolecules (polypeptides, fucoxanthin, sterols, fatty acids, polysaccharides, and polyphenols) from diatoms—A review. *Int. J. Biol. Macromol.* **2021**, *171*, 398–413. [CrossRef]
14. Grzesik, M.; Romanowska-Duda, Z.; Piotrowski, K.; Janas, R. Diatoms (Bacillariophyceae) as an effective base of a new generation of organic fertilizers. *Przemys Chem.* **2015**, *94*, 391–396. [CrossRef]
15. Hedayatkhah, A.; Cretoiu, M.S.; Emtiazi, G.; Stal, L.J.; Bolhuis, H. Bioremediation of chromium contaminated water by diatoms with concomitant lipid accumulation for biofuel production. *J. Environ. Manag.* **2018**, *227*, 313–320. [CrossRef] [PubMed]
16. Mojiri, A.; Baharlooeian, M.; Zahed, M.A. The Potential of *Chaetoceros muelleri* in Bioremediation of Antibiotics: Performance and Optimization. *Int. J. Environ. Res. Public Health* **2021**, *18*, 977. [CrossRef]
17. Lomora, M.; Shumate, D.; Rahman, A.A.; Pandit, A. Therapeutic applications of phytoplankton, with an emphasis on diatoms and coccolithophores. *Adv. Ther.* **2019**, *2*, 1800099. [CrossRef]
18. Uthappa, U.T.; Brahmkhatri, V.; Sriram, G.; Jung, H.Y.; Yu, J.; Kurkuri, N.; Aminabhavi, T.M.; Altalhi, T.; Neelgund, G.M.; Kurkuri, M.D. Nature engineered diatom biosilica as drug delivery systems. *J. Control. Release* **2018**, *281*, 70–83. [CrossRef]
19. Uthappa, U.T.; Sriram, G.; Arvind, O.R.; Kumar, S.; Ho Young, J.; Neelgund, G.M.; Losic, D.; Kurkuri, M.D. Engineering MIL-100(Fe) on 3D porous natural diatoms as a versatile high performing platform for controlled isoniazid drug release, Fenton's catalysis for malachite green dye degradation and environmental adsorbents for Pb^{2+} removal and dyes. *Appl. Surf. Sci.* **2020**, *528*, 146974. [CrossRef]
20. Ruggiero, I.; Terracciano, M.; Martucci, N.M.; De Stefano, L.; Migliaccio, N.; Tate, R.; Rendina, I.; Arcari, P.; Lamberti, A.; Rea, I. Diatomite silica nanoparticles for drug delivery. *Nanoscale Res. Lett.* **2014**, *9*, 7. [CrossRef]
21. Dalgic, A.D.; Atila, D.; Karatas, A.; Tezcaner, A.; Keskin, D. Diatom shell incorporated PHBV/PCL-pullulan co-electrospun scaffold for bone tissue engineering. *Mater. Sci. Eng. C Mater. Biol. Appl.* **2019**, *100*, 735–746. [CrossRef] [PubMed]
22. Kong, X.; Squire, K.; Li, E.; LeDuff, P.; Rorrer, G.L.; Tang, S.; Chen, B.; McKay, C.P.; Navarro-Gonzalez, R.; Wang, A.X. Chemical and Biological Sensing Using Diatom Photonic Crystal Biosilica With In-Situ Growth Plasmonic Nanoparticles. *IEEE Trans. Nanobiosci.* **2016**, *15*, 828–834. [CrossRef] [PubMed]
23. Albert, K.; Huang, X.C.; Hsu, H.Y. Bio-templated silica composites for next-generation biomedical applications. *Adv. Colloid. Interface Sci.* **2017**, *249*, 272–289. [CrossRef] [PubMed]
24. Terracciano, M.; De Stefano, L.; Rea, I. Diatoms green nanotechnology for biosilica-based drug delivery systems. *Pharmaceutics* **2018**, *10*, 242. [CrossRef] [PubMed]

25. Delasoie, J.; Zobi, F. Natural diatom biosilica as microshuttles in drug delivery systems. *Pharmaceutics* **2019**, *11*, 537. [CrossRef]
26. Tramontano, C.; Chianese, G.; Terracciano, M.; de Stefano, L.; Rea, I. Nanostructured biosilica of diatoms: From water world to biomedical applications. *Appl. Sci.* **2020**, *10*, 6811. [CrossRef]
27. Lewin, J.C. Silicon metabolism in diatoms. III. Respiration and silicon uptake in *Navicula pelliculosa*. *J. Gen. Physiol.* **1955**, *39*, 1–10. [CrossRef]
28. Annenkov, V.V.; Danilovtseva, E.N.; Pal'shin, V.A.; Ol'ga, N.V.; Zelinskiy, S.N.; Krishnan, U.M. Silicic acid condensation under the influence of water-soluble polymers: From biology to new materials. *RSC Adv.* **2017**, *7*, 20995–21027. [CrossRef]
29. Thamatrakoln, K.; Hildebrand, M. Silicon uptake in diatoms revisited: A model for saturable and nonsaturable uptake kinetics and the role of silicon transporters. *Plant Physiol.* **2008**, *146*, 1397–1407. [CrossRef]
30. Hildebrand, M.; Dahlin, K.; Volcani, B.E. Characterization of a silicon transporter gene family in *Cylindrotheca fusiformis*: Sequences, expression analysis, and identification of homologs in other diatoms. *Mol. Gen. Genet.* **1998**, *260*, 480–486. [CrossRef]
31. Hildebrand, M. Biological processing of nanostructured silica in diatoms. *Prog. Org. Coat.* **2003**, *47*, 256–266. [CrossRef]
32. Marchenkov, A.M.; Petrova, D.P.; Morozov, A.A.; Zakharova, Y.R.; Grachev, M.A.; Bondar, A.A. A family of silicon transporter structural genes in a pennate diatom *Synedra ulna* subsp. *danica* (Kutz.) Skabitsch. *PLoS ONE* **2018**, *13*, e0203161. [CrossRef]
33. Hildebrand, M.; Volcani, B.E.; Gassmann, W.; Schroeder, J.I. A gene family of silicon transporters. *Nature* **1997**, *385*, 688–689. [CrossRef]
34. Durkin, C.A.; Koester, J.A.; Bender, S.J.; Armbrust, E.V. The evolution of silicon transporters in diatoms. *J. Phycol.* **2016**, *52*, 716–731. [CrossRef]
35. Brzezinski, M.A.; Conley, D.J. Silicon deposition during the cell-cycle of *Thalassiosira weissflogii* (Bacillariophyceae) determined using dual Rhodamine-123 and propidium iodide staining. *J. Phycol.* **1994**, *30*, 45–55. [CrossRef]
36. Taylor, N.J. Silica incorporation in the diatom *Cosinodiscus granii* as affected by light intensity. *Br. Phycol. J.* **1985**, *20*, 365–374. [CrossRef]
37. Martin-Jezequel, V.; Hildebrand, M.; Brzezinski, M.A. Silicon metabolism in diatoms: Implications for growth. *J. Phycol.* **2000**, *36*, 821–840. [CrossRef]
38. Hildebrand, M. Diatoms, Biomineralization Processes, and Genomics. *Chem. Rev.* **2008**, *108*, 4855–4874. [CrossRef]
39. Crawford, R.; Schmid, A.M. Ultrastructure of silica deposition. In *Biomineralization in Lower Plants and Animals*; Leadbeater, B.S., Riding, R., Eds.; Oxford University Press: Oxford, UK, 1986; Volume 30, pp. 291–314.
40. Schmidt, A.-M. Aspects of morphogenesis and function of diatom cell walls with implications for taxonomy. *Protoplasma* **1994**, *181*, 43–60. [CrossRef]
41. Li, C.W.; Volcani, B.E. Aspects of silicification in wall morphogenesis of diatoms. *Philos. Trans. R. Soc. Lond. Ser. B Biol. Sci.* **1984**, *304*, 519–528. [CrossRef]
42. Tesson, B.; Hildebrand, M. Extensive and intimate association of the cytoskeleton with forming silica in diatoms: Control over patterning on the meso- and micro-scale. *PLoS ONE* **2010**, *5*. [CrossRef]
43. Sumper, M.; Kroger, N. Silica formation in diatoms: The function of long-chain polyamines and silaffins. *J. Mater. Chem.* **2004**, *14*, 2059–2065. [CrossRef]
44. Pahl, S.L.; Lewis, D.M.; Chen, F.; King, K.D. Heterotrophic growth and nutritional aspects of the diatom *Cyclotella cryptica* (Bacillariophyceae): Effect of some environmental factors. *J. Biosci. Bioeng.* **2010**, *109*, 235–239. [CrossRef]
45. Pahl, S.L.; Lewis, D.M.; Chen, F.; King, K.D. Growth dynamics and the proximate biochemical composition and fatty acid profile of the heterotrophically grown diatom *Cyclotella Cryptica*. *J. Appl. Phycol.* **2010**, *22*, 165–171. [CrossRef] [PubMed]
46. Khan, M.; Karmakar, R.; Das, B.; Diba, F.; Razu, M.H. Recent advances in microalgal biotechnology. In *Heterotrophic Growth of Micro Algae*; Jin, L., Zheng, S., Henri, G., Eds.; OMICS Group eBooks: Foster City, CA, USA, 2016; pp. 1–18.
47. Mao, X.M.; Chen, S.H.Y.; Lu, X.; Yu, J.F.; Liu, B. High silicate concentration facilitates fucoxanthin and eicosapentaenoic acid (EPA) production under heterotrophic condition in the marine diatom *Nitzschia laevis*. *Algal Res. Biomass Biofuels Bioprod.* **2020**, *52*. [CrossRef]
48. Chen, G.Q.; Chen, F. Growing phototrophic cells without light. *Biotechnol. Lett.* **2006**, *28*, 607–616. [CrossRef]
49. Morales-Sanchez, D.; Martinez-Rodriguez, O.A.; Kyndt, J.; Martinez, A. Heterotrophic growth of microalgae: Metabolic aspects. *World J. Microbiol. Biotechnol.* **2015**, *31*, 1–9. [CrossRef]
50. D'Ippolito, G.; Sardo, A.; Paris, D.; Vella, F.M.; Adelfi, M.G.; Botte, P.; Gallo, C.; Fontana, A. Potential of lipid metabolism in marine diatoms for biofuel production. *Biotechnol. Biofuels* **2015**, *8*, 28. [CrossRef] [PubMed]
51. Botte, P.; D'Ippolito, G.; Gallo, C.; Sardo, A.; Fontana, A. Combined exploitation of CO_2 and nutrient replenishment for increasing biomass and lipid productivity of the marine diatoms *Thalassiosira weissflogii* and *Cyclotella cryptica*. *J. Appl. Phycol.* **2017**, *30*, 243–251. [CrossRef]
52. Orefice, I.; Musella, M.; Smerilli, A.; Sansone, C.; Chandrasekaran, R.; Corato, F.; Brunet, C. Role of nutrient concentrations and water movement on diatom's productivity in culture. *Sci. Rep.* **2019**, *9*, 1479. [CrossRef]
53. Lutyński, M.; Sakiewicz, P.; Lutyńska, S. Characterization of diatomaceous earth and halloysite resources of poland. *Minerals* **2019**, *9*, 670. [CrossRef]
54. Qin, T.; Gutu, T.; Jiao, J.; Chang, C.H.; Rorrer, G.L. Photoluminescence of silica nanostructures from bioreactor culture of marine diatom *Nitzschia frustulum*. *J. Nanosci. Nanotechnol.* **2008**, *8*, 2392–2398. [CrossRef]

55. Wang, Y.; Cai, J.; Jiang, Y.G.; Jiang, X.G.; Zhang, D.Y. Preparation of biosilica structures from frustules of diatoms and their applications: Current state and perspectives. *Appl. Microbiol. Biotechnol.* **2013**, *97*, 453–460. [CrossRef]
56. Qi, Y.R.; Wang, X.; Cheng, J.J. Preparation and characteristics of biosilica derived from marine diatom biomass of *Nitzschia closterium* and *Thalassiosira*. *Chin. J. Oceanol. Limnol.* **2017**, *35*, 668–680. [CrossRef]
57. Horn, M.G.; Robinson, R.S.; Rynearson, T.A.; Sigman, D.M. Nitrogen isotopic relationship between diatom-bound and bulk organic matter of cultured polar diatoms. *Paleoceanography* **2011**, *26*. [CrossRef]
58. Mejia, L.M.; Isensee, K.; Mendez-Vicente, A.; Pisonero, J.; Shimizu, N.; Gonzalez, C.; Monteleone, B.; Stoll, H. B content and Si/C ratios from cultured diatoms (*Thalassiosira pseudonana* and *Thalassiosira weissflogii*): Relationship to seawater pH and diatom carbon acquisition. *Geochim. Cosmochim. Acta* **2013**, *123*, 322–337. [CrossRef]
59. Jiang, W.K.; Luo, S.P.; Liu, P.W.; Deng, X.Y.; Jing, Y.; Bai, C.Y.; Li, J.B. Purification of biosilica from living diatoms by a two-step acid cleaning and baking method. *J. Appl. Phycol.* **2014**, *26*, 1511–1518. [CrossRef]
60. Umemura, K.; Noguchi, Y.; Ichinose, T.; Hirose, Y.; Kuroda, R.; Mayama, S. Diatom Cells Grown and Baked on a Functionalized Mica Surface. *J. Biol. Phys.* **2008**, *34*, 189–196. [CrossRef]
61. Watanabe, T.; Kodama, Y.; Mayama, S. Application of a novel cleaning method using low-temperature plasma on tidal flat diatoms with heterovalvy or delicate frustule structure. *Proc. Acad. Nat. Sci. Phila.* **2010**, *160*, 83–87. [CrossRef]
62. Saad, E.M.; Pickering, R.A.; Shoji, K.; Hossain, M.I.; Glover, T.G.; Krause, J.W.; Tang, Y.Z. Effect of cleaning methods on the dissolution of diatom frustules. *Mar. Chem.* **2020**, *224*. [CrossRef]
63. Jeffryes, C.; Solanki, R.; Rangineni, Y.; Wang, W.; Chang, C.H.; Rorrer, G.L. Electroluminescence and photoluminescence from nanostructured diatom frustules containing metabolically inserted germanium. *Adv. Mater.* **2008**, *20*, 2633–2637. [CrossRef]
64. Townley, H.E.; Parker, A.R.; White-Cooper, H. Exploitation of diatom frustules for nanotechnology: Tethering active biomolecules. *Adv. Funct. Mater.* **2008**, *18*, 369–374. [CrossRef]
65. Abramson, L.; Wirick, S.; Lee, C.; Jacobsen, C.; Brandes, J.A. The use of soft X-ray spectromicroscopy to investigate the distribution and composition of organic matter in a diatom frustule and a biomimetic analog. *Deep Sea Res. Part II Top. Stud. Oceanogr.* **2009**, *56*, 1369–1380. [CrossRef]
66. Lin, K.C.; Kunduru, V.; Bothara, M.; Rege, K.; Prasad, S.; Ramakrishna, B.L. Biogenic nanoporous silica-based sensor for enhanced electrochemical detection of cardiovascular biomarkers proteins. *Biosens. Bioelectron.* **2010**, *25*, 2336–2342. [CrossRef] [PubMed]
67. Bariana, M.; Aw, M.S.; Kurkuri, M.; Losic, D. Tuning drug loading and release properties of diatom silica microparticles by surface modifications. *Int. J. Pharm.* **2013**, *443*, 230–241. [CrossRef]
68. Van Eynde, E.; Lenaerts, B.; Tytgat, T.; Verbruggen, S.W.; Hauchecorne, B.; Blust, R.; Lenaerts, S. Effect of pretreatment and temperature on the properties of *Pinnularia* biosilica frustules. *RSC Adv.* **2014**, *4*, 56200–56206. [CrossRef]
69. Lettieri, S.; Setaro, A.; De Stefano, L.; De Stefano, M.; Maddalena, P. The gas-detection properties of light-emitting diatoms. *Adv. Funct. Mater.* **2008**, *18*, 1257–1264. [CrossRef]
70. De Stefano, L.; Rendina, I.; De Stefano, M.; Bismuto, A.; Maddalena, P. Marine diatoms as optical chemical sensors. *Appl. Phys. Lett.* **2005**, *87*. [CrossRef]
71. De Stefano, L.; Rotiroti, L.; De Stefano, M.; Lamberti, A.; Lettieri, S.; Setaro, A.; Maddalena, P. Marine diatoms as optical biosensors. *Biosens. Bioelectron.* **2009**, *24*, 1580–1584. [CrossRef]
72. Wang, Y.; Zhang, D.Y.; Cai, J.; Pan, J.F.; Chen, M.L.; Li, A.B.; Jiang, Y.G. Biosilica structures obtained from *Nitzschia, Ditylum, Skeletonema, and Coscinodiscus* diatom by a filtration-aided acid cleaning method. *Appl. Microbiol. Biotechnol.* **2012**, *95*, 1165–1178. [CrossRef]
73. Delalat, B.; Sheppard, V.C.; Rasi Ghaemi, S.; Rao, S.; Prestidge, C.A.; McPhee, G.; Rogers, M.L.; Donoghue, J.F.; Pillay, V.; Johns, T.G.; et al. Targeted drug delivery using genetically engineered diatom biosilica. *Nat. Commun.* **2015**, *6*, 8791. [CrossRef] [PubMed]
74. Abdelhamid, M.A.A.; Pack, S.P. Biomimetic and bioinspired silicifications: Recent advances for biomaterial design and applications. *Acta Biomater.* **2021**, *120*, 38–56. [CrossRef] [PubMed]
75. Vasani, R.B.; Losic, D.; Cavallaro, A.; Voelcker, N.H. Fabrication of stimulus-responsive diatom biosilica microcapsules for antibiotic drug delivery. *J. Mater. Chem. B* **2015**, *3*, 4325–4329. [CrossRef] [PubMed]
76. Patel, P.; Hanini, A.; Shah, A.; Patel, D.; Patel, S.; Bhatt, P.; Pathak, V.Y. Surface modification of nanoparticles for targeted drug delivery. In *Surface Modification of Nanoparticles for Targeted Drug Delivery*; Pathak, V.Y., Ed.; Springer: Berlin/Heidelberg, Germany, 2019.
77. Bao, Z.; Weatherspoon, M.R.; Shian, S.; Cai, Y.; Graham, P.D.; Allan, S.M.; Ahmad, G.; Dickerson, M.B.; Church, B.C.; Kang, Z.; et al. Chemical reduction of three-dimensional silica micro-assemblies into microporous silicon replicas. *Nature* **2007**, *446*, 172–175. [CrossRef] [PubMed]
78. Losic, D.; Yu, Y.; Aw, M.S.; Simovic, S.; Thierry, B.; Addai-Mensah, J. Surface functionalisation of diatoms with dopamine modified iron-oxide nanoparticles: Toward magnetically guided drug microcarriers with biologically derived morphologies. *Chem. Commun.* **2010**, *46*, 6323–6325. [CrossRef]
79. Jantschke, A.; Fischer, C.; Hensel, R.; Braun, H.G.; Brunner, E. Directed assembly of nanoparticles to isolated diatom valves using the non-wetting characteristics after pyrolysis. *Nanoscale* **2014**, *6*, 11637–11645. [CrossRef]
80. Diab, M.; Mokari, T. Bioinspired hierarchical porous structures for engineering advanced functional inorganic materials. *Adv. Mater.* **2018**, *30*, e1706349. [CrossRef]

81. Aw, M.S.; Simovic, S.; Yu, Y.; Addai-Mensah, J.; Losic, D. Porous silica microshells from diatoms as biocarrier for drug delivery applications. *Powder Technol.* **2012**, *223*, 52–58. [CrossRef]
82. Kumeria, T.; Bariana, M.; Altalhi, T.; Kurkuri, M.; Gibson, C.T.; Yang, W.; Losic, D. Graphene oxide decorated diatom silica particles as new nano-hybrids: Towards smart natural drug microcarriers. *J. Mater. Chem. B* **2013**, *1*, 6302–6311. [CrossRef]
83. Delasoie, J.; Rossier, J.; Haeni, L.; Rothen-Rutishauser, B.; Zobi, F. Slow-targeted release of a ruthenium anticancer agent from vitamin B12 functionalized marine diatom microalgae. *Dalton Trans.* **2018**, *47*, 17221–17232. [CrossRef]
84. Maher, S.; Santos, A.; Kumeria, T.; Kaur, G.; Lambert, M.; Forward, P.; Evdokiou, A.; Losic, D. Multifunctional microspherical magnetic and pH responsive carriers for combination anticancer therapy engineered by droplet-based microfluidics. *J. Mater. Chem. B* **2017**, *5*, 4097–4109. [CrossRef] [PubMed]
85. Milovic, M.; Simovic, S.; Losic, D.; Dashevskiy, A.; Ibric, S. Solid self-emulsifying phospholipid suspension (SSEPS) with diatom as a drug carrier. *Eur. J. Pharm. Sci.* **2014**, *63*, 226–232. [CrossRef]
86. Singh, R.; Khan, M.J.; Rane, J.; Gajbhiye, A.; Vinayak, V.; Joshi, K.B. Biofabrication of Diatom Surface by Tyrosine-Metal Complexes: Smart Microcontainers to Inhibit Bacterial Growth. *Chemistryselect* **2020**, *5*, 3091–3097. [CrossRef]
87. Esfandyari, J.; Shojaedin-Givi, B.; Hashemzadeh, H.; Mozafari-Nia, M.; Vaezi, Z.; Naderi-Manesh, H. Capture and detection of rare cancer cells in blood by intrinsic fluorescence of a novel functionalized diatom. *Photodiagn. Photodyn. Ther.* **2020**, *30*, 101753. [CrossRef] [PubMed]
88. De Stefano, L.; Lamberti, A.; Rotiroti, L.; De Stefano, M. Interfacing the nanostructured biosilica microshells of the marine diatom *Coscinodiscus wailesii* with biological matter. *Acta Biomater.* **2008**, *4*, 126–130. [CrossRef]
89. Gale, D.K.; Gutu, T.; Jiao, J.; Chang, C.-H.; Rorrer, G.L. Photoluminescence Detection of Biomolecules by Antibody-Functionalized Diatom Biosilica. *Adv. Funct. Mater.* **2009**, *19*, 926–933. [CrossRef]
90. Cicco, S.R.; Vona, D.; Gristina, R.; Sardella, E.; Ragni, R.; Lo Presti, M.; Farinola, G.M. Biosilica from Living Diatoms: Investigations on Biocompatibility of Bare and Chemically Modified *Thalassiosira weissflogii* Silica Shells. *Bioengineering* **2016**, *3*, 35. [CrossRef]
91. Aw, M.S.; Bariana, M.; Yu, Y.; Addai-Mensah, J.; Losic, D. Surface-functionalized diatom microcapsules for drug delivery of water-insoluble drugs. *J. Biomater. Appl.* **2013**, *28*, 163–174. [CrossRef] [PubMed]
92. Martucci, N.M.; Migliaccio, N.; Ruggiero, I.; Albano, F.; Cali, G.; Romano, S.; Terracciano, M.; Rea, I.; Arcari, P.; Lamberti, A. Nanoparticle-based strategy for personalized B-cell lymphoma therapy. *Int. J. Nanomed.* **2016**, *11*, 6089–6101. [CrossRef]
93. Rea, I.; Martucci, N.M.; De Stefano, L.; Ruggiero, I.; Terracciano, M.; Dardano, P.; Migliaccio, N.; Arcari, P.; Tate, R.; Rendina, I.; et al. Diatomite biosilica nanocarriers for siRNA transport inside cancer cells. *Biochim. Biophys. Acta Gen. Subj.* **2014**, *1840*, 3393–3403. [CrossRef]
94. Zhang, H.; Shahbazi, M.A.; Makila, E.M.; da Silva, T.H.; Reis, R.L.; Salonen, J.J.; Hirvonen, J.T.; Santos, H.A. Diatom silica microparticles for sustained release and permeation enhancement following oral delivery of prednisone and mesalamine. *Biomaterials* **2013**, *34*, 9210–9219. [CrossRef] [PubMed]
95. Terracciano, M.; Shahbazi, M.A.; Correia, A.; Rea, I.; Lamberti, A.; De Stefano, L.; Santos, H.A. Surface bioengineering of diatomite based nanovectors for efficient intracellular uptake and drug delivery. *Nanoscale* **2015**, *7*, 20063–20074. [CrossRef]
96. Amoda, A.; Borkiewicz, L.; Rivero-Müller, A.; Alam, P. Sintered nanoporous biosilica diatom frustules as high efficiency cell-growth and bone-mineralisation platforms. *Mater. Today Commun.* **2020**, *24*, 100923. [CrossRef]
97. Terracciano, M.; De Stefano, L.; Tortiglione, C.; Tino, A.; Rea, I. In Vivo Toxicity Assessment of Hybrid Diatomite Nanovectors Using *Hydra vulgaris* as a Model System. *Adv. Biosyst.* **2019**, *3*, e1800247. [CrossRef] [PubMed]
98. Poulsen, N.; Berne, C.; Spain, J.; Kroger, N. Silica immobilization of an enzyme through genetic engineering of the diatom *Thalassiosira pseudonana*. *Angew. Chem. Int. Ed.* **2007**, *46*, 1843–1846. [CrossRef]
99. Kumari, E.; Görlich, S.; Poulsen, N.; Kröger, N. Genetically Programmed Regioselective Immobilization of Enzymes in Biosilica Microparticles. *Adv. Funct. Mater.* **2020**, *30*, 2000442. [CrossRef]
100. Ford, N.R.; Hecht, K.A.; Hu, D.H.; Orr, G.; Xiong, Y.J.; Squier, T.C.; Rorrer, G.L.; Roesijadi, G. Antigen Binding and Site-Directed Labeling of Biosilica-Immobilized Fusion Proteins Expressed in Diatoms. *ACS Synth. Biol.* **2016**, *5*, 193–199. [CrossRef]
101. Vona, D.; Urbano, L.; Bonifacio, M.A.; De Giglio, E.; Cometa, S.; Mattioli-Belmonte, M.; Palumbo, F.; Ragni, R.; Cicco, S.R.; Farinola, G.M. Data from two different culture conditions of *Thalassiosira weissflogii* diatom and from cleaning procedures for obtaining monodisperse nanostructured biosilica. *Data Brief* **2016**, *8*, 312–319. [CrossRef]
102. Saxena, A.; Tiwari, A.; Kaushik, R.; Iqbal, H.M.N.; Parra-Saldivar, R. Diatoms recovery from wastewater: Overview from an ecological and economic perspective. *J. Water Process Eng.* **2021**, *39*. [CrossRef]
103. Diab, R.; Canilho, N.; Pavel, I.A.; Haffner, F.B.; Girardon, M.; Pasc, A. Silica-based systems for oral delivery of drugs, macromolecules and cells. *Adv. Colloid Interface Sci.* **2017**, *249*, 346–362. [CrossRef]
104. Cauda, V.; Schlossbauer, A.; Bein, T. Bio-degradation study of colloidal mesoporous silica nanoparticles: Effect of surface functionalization with organo-silanes and poly(ethylene glycol). *Microporous Mesoporous Mater.* **2010**, *132*, 60–71. [CrossRef]
105. Anderson, S.H.C.; Elliott, H.; Wallis, D.J.; Canham, L.T.; Powell, J.J. Dissolution of different forms of partially porous silicon wafers under simulated physiological conditions. *Phys. Status Solidi A Appl. Mater. Sci.* **2003**, *197*, 331–335. [CrossRef]
106. Popplewell, J.F.; King, S.J.; Day, J.P.; Ackrill, P.; Fifield, L.K.; Cresswell, R.G.; Di Tada, M.L.; Liu, K. Kinetics of uptake and elimination of silicic acid by a human subject: A novel application of Si-32 and accelerator mass spectrometry. *J. Inorg. Biochem.* **1998**, *69*, 177–180. [CrossRef]

107. Zhang, Y.; Hsu, B.Y.W.; Ren, C.L.; Li, X.; Wang, J. Silica-based nanocapsules: Synthesis, structure control and biomedical applications. *Chem. Soc. Rev.* **2015**, *44*, 315–335. [CrossRef]
108. Wallace, A.K.; Chanut, N.; Voigt, C.A. Silica Nanostructures Produced Using Diatom Peptides with Designed Post-Translational Modifications. *Adv. Funct. Mater.* **2020**, *30*. [CrossRef]
109. Rogato, A.; De Tommasi, E. Physical, Chemical, and Genetic Techniques for Diatom Frustule Modification: Applications in Nanotechnology. *Appl. Sci.* **2020**, *10*, 8738. [CrossRef]
110. Sun, Z.; Zhou, Z.; Li, J.; Feng, C.; Chen, X.; Zhang, Y. Industrialized System for Producing Bio-Silicon, Has Centrifugal Shaft That is Connected to Barrel Structure of Inner Barrel, and Flocculation Tank Whose Outlet is Connected to Algae-Liquid Separation Tank. China Patent CN110616139-A, 27 December 2019. CN110616139-B, 28 August 2020.
111. Kaya, M.; Bilican, I.; Mujtaba, M.; Sargin, I.; Haskoylu, M.E.; Oner, E.T.; Zheng, K.; Boccaccini, A.R.; Cansaran-Duman, D.; Onses, M.S.; et al. Sponge-derived natural bioactive glass microspheres with self-assembled surface channel arrays opening into a hollow core for bone tissue and controlled drug release applications. *Chem. Eng. J.* **2021**, *407*. [CrossRef]
112. Granito, R.N.; Custodio, M.R.; Renno, A.C.M. Natural marine sponges for bone tissue engineering: The state of art and future perspectives. *J. Biomed. Mater. Res. Part B Appl. Biomater.* **2017**, *105*, 1717–1727. [CrossRef]
113. Miricioiu, M.G.; Niculescu, V.C.; Filote, C.; Raboaca, M.S.; Nechifor, G. Coal Fly Ash Derived Silica Nanomaterial for MMMs-Application in CO_2/CH_4 Separation. *Membranes* **2021**, *11*, 78. [CrossRef]
114. Miricioiu, M.G.; Niculescu, V.C. Fly ash, from recycling to potential raw material for mesoporous silica synthesis. *Nanomaterials* **2020**, *10*, 474. [CrossRef]
115. Jeffryes, C.; Agathos, S.N.; Rorrer, G. Biogenic nanomaterials from photosynthetic microorganisms. *Curr. Opin. Biotechnol.* **2015**, *33*, 23–31. [CrossRef] [PubMed]
116. Castillo, R.R.; Vallet-Regi, M. Functional mesoporous silica nanocomposites: Biomedical applications and biosafety. *Int. J. Mol. Sci.* **2019**, *20*, 929. [CrossRef] [PubMed]

Review

Advances in Purpurin 18 Research: On Cancer Therapy

Vladimíra Pavlíčková [1], Jan Škubník [1], Michal Jurášek [2] and Silvie Rimpelová [1,*]

[1] Department of Biochemistry and Microbiology, University of Chemistry and Technology Prague, Technická 5, 166 28 Prague 6, Czech Republic; vladimira.pavlickova@vscht.cz (V.P.); jan.skubnik@vscht.cz (J.Š.)
[2] Department of Chemistry of Natural Compounds, University of Chemistry and Technology Prague, Technická 5, 166 28 Prague 6, Czech Republic; michal.jurasek@vscht.cz
* Correspondence: silvie.rimpelova@vscht.cz

Abstract: How to make cancer treatment more efficient and enhance the patient's outcome? By multimodal therapy, theranostics, or personalized medicine? These are questions asked by scientists and doctors worldwide. However, finding new unique approaches and options for cancer treatment as well as new selective therapeutics is very challenging. More frequently, researchers "go back in time" and use already known and well-described compounds/drugs, the structure of which further derivatize to "improve" their properties, extend the use of existing drugs to new indications, or even to obtain a completely novel drug. Natural substances, especially marine products, are a great inspiration in the discovery and development of novel anticancer drugs. These can be used in many modern approaches, either as photo- and sonosensitizers in photodynamic and sonodynamic cancer therapy, respectively, or in tumor imaging and diagnosis. This review is focused on a very potent natural product, the chlorophyll metabolite purpurin 18, and its derivatives, which is well suitable for all the mentioned applications. Purpurin 18 can be easily isolated from green plants of all kinds ranging from seaweed to spinach leaves and, thus, it presents an economically feasible source for a very promising anticancer drug.

Keywords: cancer therapy; gold nanoparticles; marine products; natural products; photosensitizer; purpurin 18; reactive oxygen species; seaweed

Citation: Pavlíčková, V.; Škubník, J.; Jurášek, M.; Rimpelová, S. Advances in Purpurin 18 Research: On Cancer Therapy. *Appl. Sci.* **2021**, *11*, 2254. https://doi.org/10.3390/app11052254

Academic Editor: Genoveffa Nuzzo

Received: 16 February 2021
Accepted: 28 February 2021
Published: 4 March 2021

1. Introduction

Cancer and the lack of an effective treatment have been global problems for many years. Current treatment approaches are based on decades-old methods such as tumor resection, radiotherapy, and classic chemotherapy, which are not always effective and, thus, cancer remains to be the leading cause of death worldwide [1]. Significant problems of current anticancer therapy are especially the low selectivity of anticancer drugs and their high systemic toxicity, which leads to damage of healthy cells. A severe obstacle in cancer treatment is also the frequent emergence of resistance to individual chemotherapeutics, whether it is a natural resistance or acquired, for example, due to insufficient drug concentration at the tumor site. Modern strategies for cancer treatment include early diagnosis of the disease as well as effective targeted treatment with minimal side effects. To the long-known methods with such benefits belongs photodynamic therapy (PDT), which is based on the application of a photosensitive substance (PS), which after excitation by light and reaction with cellular oxygen generates cytotoxic reactive oxygen species (ROS) [2]. Very similar methods closely related to PDT are photothermal therapy (PTT) and sonodynamic therapy (SDT). In PTT, the light energy absorbed by a PS is converted into thermal energy. The generated heat then warms the target tumor cells, which leads to their damage and death. In the case of SDT, instead of light, ultrasound waves are applied to activate the drug used, a so-called sonosensitizer (ST). Then, the activated ST eliminates the target cells by a similar mechanism to that of a PS. In addition to SDT, another method utilizes acoustic ultrasonic waves—photoacoustic imaging (PAI). However, as for PAI, the ultrasound waves do not serve as drug activators but as a measurement output. In the case of PAI, the drug is activated by a pulsed laser, and based on the energy conversion, ultrasonic waves are subsequently generated, which are monitored to form an image. Owing to the high resolution of PAI and the possibility of deep penetration into the tissues, this method is a promising technique, especially in terms of tumor diagnosis. However, for all these methods, very specific compounds, e.g., of natural origin, must be sought.

Natural compounds for cancer therapies have often a plant origin, but many of them come also from marine organisms and seaweed [3]. Many substances isolated from seaweed have been shown to have significant health benefits and are thus often used in the cosmetics or food and pharmaceutical industries. Seaweed can be classified as an important source of polyunsaturated fatty acids, vitamins, enzymes but also natural pigments such as chlorophylls, carotenoids, and quinones [4–8]. Evidence suggests that the most important are green pigments—chlorophylls. These occur in brown, red, and green seaweed (*Phaeophyceae*, *Rhodophyceae*, and *Chlorophyceae*) [9]. Due to the high content of chlorophyll (Chl; 565 to 2000 $mg \cdot kg^{-1}$ of dried biomass) in seaweed [8] and undemanding cultivation methods on a large scale, seaweed is a favorable source for Chl production. Dried seaweed biomass can be used to extract Chl utilizing organic solvents or supercritical extraction [10]. The Chl thus obtained can then be chemically modified to give its derivatives and metabolites, one of which is purpurin 18 (pu18), which is well known for its unique properties and great potential in anticancer therapy.

This review aims to detail the most recent advances in pu18 anticancer research, with emphasis especially on PDT, PTT, and PAI. Moreover, biosynthesis of pu18 and novel anticancer derivatives of pu18 are also discussed.

2. From Plant Dyes to Medicines

Chl is perhaps the most famous green pigment that each of us comes across daily. The most common and most important in terms of photosynthesis is blue-green chlorophyll *a* (Chl*a*), which absorbs light energy at wavelengths with a maximum of 660–665 nm [10,11]. Other types are yellow-green chlorophyll *b* (Chl*b*, maximum absorbance at 642–652 nm), which occurs in green seaweed, and chlorophyll *c* with a maximum absorbance of 447–452 nm. The fourth type, chlorophyll *d* absorbs at far-red wavelengths around 710 nm and occurs in cyanobacteria and red seaweed [12]. Chl plays a key role in plant photosynthesis and its decay is an important catabolic process of leaf aging and fruit ripening. However, the biochemical processes by which Chl is degraded have been unknown for many years, and it was not until the identification of the major metabolites that the whole process was elucidated. It is now known that in the first phase, enzymatic reactions of color pigments take place. The pigments serve as substrates for the formation of a colorless, blue fluorescent decomposition product, called primary fluorescent Chl catabolite. One of the key steps in this metabolic pathway is the enzymatic conversion of Chl*b* to Chl*a*, also known as the Chl cycle [13,14]. The central magnesium atom is further removed from the Chl*a* molecule by magnesium dechelatase [EC 4.99.1.10] to form pheophytin *a*, which passes into another colored product pheophorbide *a* (Pb*a*). Through other enzyme-catalyzed reactions, Pb*a* is converted to red Chl catabolite and further to primary fluorescent Chl catabolite. Chl degradation usually results in non-enzymatic isomerization of modified fluorescent Chl catabolites to their respective nonfluorescent Chl catabolite, which is transported into plant cell vacuoles for further processing (Figure 1) [15].

As the structural formulas are shown in Figure 1, it is apparent that Chl is a porphyrin with a cyclic tetrapyrrole nucleus. Its metabolites are characterized by 18π electrons in the aromatic ring with a conjugated double bond system. This structural arrangement predetermines Chl metabolites to be PSs with the possibility of use for example in cancer therapy. Chl*a* itself is characterized by a high extinction coefficient at 661 nm [16] and good singlet oxygen production. However, what limits its further use is its low thermal stability, proneness to degradation due to light irradiation, and also low stability at acidic pH. At the same time, it was found that Chl*a* exhibits lower thermal and pH stability than

Chl*b* [17]. Therefore, from the point of view of a PS, its natural degradation products and their semi-synthetic derivatives are more interesting for potential medicinal application.

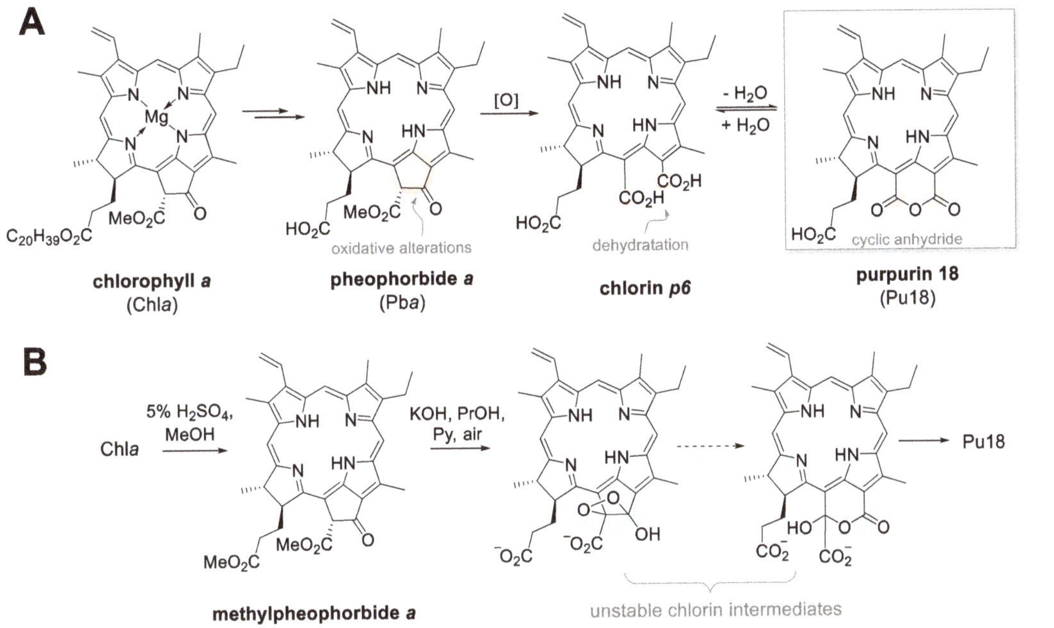

Figure 1. The biological degradation process of chlorophyll *a* leads to purpurin 18 (pu18, panel **A**) and chemical transformation of chlorophyll *a* to pu18 (panel **B**).

3. Purpurin 18 and Cancer Treatment

One of such degradation products of Chl is the already aforementioned pu18, a photosensitive chlorin substance naturally occurring in, for example, spinach leaves [18], cyanobacterium *Spirulina maxima* [19], Japanese carpet shells *Ruditapes phlippinarum* [20], brown mussel *Perna perna* [21], or in a whole series of edible seaweed [22]. It has been confirmed that pu18 can be even observed from geological samples of sediments [23]. Considering mainly its photochemical properties, pu18 represents a potentially useful substance for cancer phototherapy. What is mainly advantageous is its maximum absorption at 700 nm, as the light of such wavelength penetrates deep into the tumor tissue [24]. Unfortunately, even despite this benefit, pu18 has not been applied in clinical practice, yet, since it exhibits high hydrophobicity. Therefore, tremendous effort has been made to derivatize pu18 structure to overcome this solubility limitation. The chemical structure of pu18 offers great options for structural modifications (Figure 2). One example among many is linking pu18 with a peptide by a reaction of the carboxyl group of propionic acid of pu18 and amidogen radical of the peptide. For example, by linking pu18 with arginine-rich peptides, conjugates with higher hydrophilicity can be prepared, which, besides, exert modified aggregation properties (increased with a higher number of arginine units) and well-ordered assembly [25]. That is, indeed, another advantage of pu18. Its macrocyclic nature makes it an ideal structural unit for creating larger entities by the supramolecular assembly. This property could be used in cancer treatment and imaging (see Chapter 9), as certain peptide conjugated pu18 derivatives can create *in situ* nanofibers and, thus, prolong retention time in tumors and, thus, enable improved imaging opportunities and more effective therapeutical outcome [26]. Therapy with pu18 is, as aforementioned, based mainly on its photosensitivity. This phenomenon is utilized in diverse therapeutical approaches, mainly PDT and PTT. Pu18, as well as other PSs, is also able to absorb energy delivered into tumors by ultrasound, which is used in SDT. Both PDT and SDT might be also combined with cancer diagnostics and, thus, pu18 can act as a theranostic agent. In the following chapters, we discuss the basics of all these approaches as well as the potential therapeutic and diagnostic roles of pu18 and its derivatives in cancer.

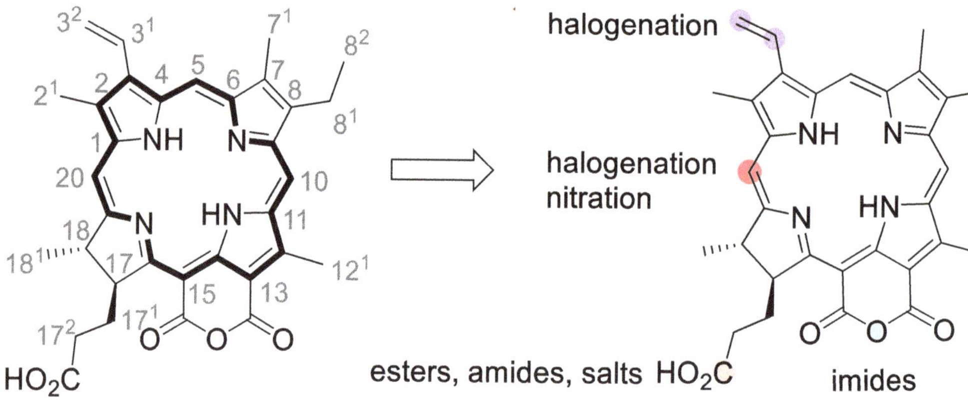

Figure 2. Numbered molecular structure of purpurin 18 (pu18) and positions of frequent modifications. The pu18 molecule is usually modified to improve its photochemical properties, for example, by halogenation (violet dots) or nitration (orange dots), or derivatized by conjugation reactions with peptides or other tumor cell-targeting molecules. The most common approaches are the use of a free carboxyl group (C-173) to form esters or amides (an orange dot) or by the derivatization of a cyclic anhydride to form imides (a blue dot).

4. Photodynamic Therapy

PDT is an attractive approach to the treatment of many types of cancer and non-cancerous diseases. This approach is based on the administration of photoactive substances, so-called PSs, which themselves (without any stimulation) do not exhibit any biological activity. They are activated by illumination with light of appropriate wavelength upon which a PS transits into an excited state. Then, the excited PS can react with molecular oxygen to form ROS and singlet oxygen. The presence of these particles in a cell leads to strong oxidative stress resulting in cell damage and cell death (Figure 3). However, ROS play a dual role in cancer treatment, since they can trigger multiple cellular cascades leading also to pro-tumorigenesis [27]. The ideal PS should meet several criteria, such as high quantum yield of singlet oxygen and absorption in the dark red or near-infrared region of the electromagnetic spectrum (650–800 nm), which ensures deeper penetration of the light into a treated tissue. On the contrary, negligible absorption of the PS in the visible part electromagnetic spectrum (400–600 nm) is desirable due to possible photosensitization of the skin [28]. Excitation and emission spectra are very important properties of PSs because human tissue contains many endogenous chromophores that also absorb light. For example, hemoglobin, myoglobin, melanin, and water absorb light in the 600–1000 nm region, which limits the biological window in this range of wavelengths [29]. Therefore, current trend is to develop novel PSs absorbing in the range of the second and third near-infrared (NIR) region of the electromagnetic spectrum (NIR-II: 1000–1350 nm, NIR-III: 1550–1870 nm), which provides the benefit of deeper tissue penetration, enhanced image contrast due to reduced optical scattering, and lower phototoxicity [30]. Related to this, it is another important property of a PS—preferred and ideally selective accumulation of the PS in the target, usually tumor, tissue. Equally important in PDT is none or marginal dark toxicity (without photoactivation) of the PS and its potential metabolites. What makes the use of PDT for cancer treatment more advantageous over conventional surgical methods are its low invasiveness and minimal side effects, which are frequently encountered in chemotherapy and radiotherapy. For this reason, PDT has been a very attractive method in cancer treatment.

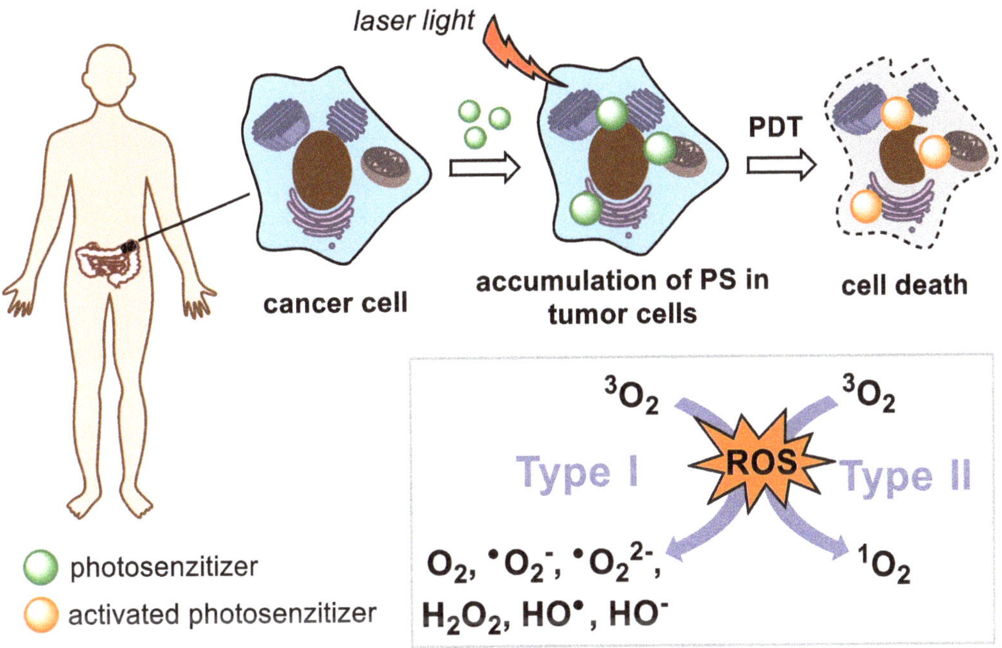

Figure 3. A general principle of photodynamic therapy (PDT). PDT is based on the action of light radiation on a photosensitive drug (PS). After receiving energy, it passes into a higher energy state. Then, it transfers the received energy to surrounding molecules and turns back to its ground state. In the energy transfer mechanism Type I, the energy is transferred to a biomolecule which reacts with $^3O_2$ to form reactive oxygen species (ROS). During the Type II transfer, there is a direct reaction of the PS with the $^3O_2$ molecule, with the formation of singlet oxygen $^1O_2$, which also belongs to the ROS.

In PDT, substances originating in natural products are often used as PSs, including the already mentioned pu18. Pu18 in connection to photoactivation has attracted scientific interest already more than thirty years ago when Hoober et al. [18] discovered that photoactivated pu18 inhibits the growth of primary human fibroblasts. Thirteen years later, after the reports that photoactivated pu18 induces apoptosis and necrosis also in human cancer cells, namely HL-60 (cells from leukemia) [31,32], the real boom in pu18 research related to cancer treatment has started.

More recently, pu18 has been tested as an effective PS for PDT treatment of commonly incurable triple-negative breast cancer [33]. The phototoxicity of pu18 was determined using a mouse mammary tumor cell line (4T1), which is used as a model of stage IV human breast cancer. 4T1 cells were used in the form of two-dimensional (2D) culture, three-dimensional (3D) spheroids, and, also, in vivo in Balb/c mice with subcutaneously implanted 4T1 cells. In the case of 4T1 2D cell cultures, significant phototoxicity of pu18 (0.5 μM) was observed after light illumination at 630 nm (35 s, 3.71 $J{\cdot}cm^{-2}$). Without photoactivation, pu18 did not exhibit any signs of toxicity up to 4 μM concentration. During the microscopic analysis of PDT-treated cells, a change in the cell shape (size reduction, rounding, vacuolation) and their separation from the surface of the culture dish was detected. Flow cytometry analysis showed that PDT treatment of 4T1 cells using 0.5 μM pu18 (0.724 $J{\cdot}cm^{-2}$) successfully induced apoptosis in 50 % of the cells. For 4T1 cells cultured as 3D spheroids, the cells were exposed to 4 μM pu18 for 24 h and were photoactivated at 630 nm (4 min, 25.2 $J{\cdot}cm^{-2}$), and further cultured for 6 days. On the second day after PDT, a decrease in the mean of 4T1 spheroids was observed compared to the control group. However, further cultivation showed that in the PDT-treated group, cell growth was not completely inhibited, since the 4T1 spheroids were further increasing their volume. However, it was slower than in the case of PDT-untreated 4T1 spheroids. The PDT effect of pu18 administered intratumorally was verified using experimental Balb/c mice with subcutaneously implanted 4T1 cells. After PDT treatment (pu18, 0.625 mg·kg^{-1}, 600 J·cm^{-2}), the tumor size was monitored every second day for two weeks. Six days after PDT, a significant decrease in the tumor volume (60%) was observed in the PDT-treated group compared to control groups, in which a constant growth of the tumors was observed. Although PDT significantly suppressed the size of the tumor, its overall disappearance did not occur here either. However, the survival time of PDT-treated mice was longer compared to the control group. The obtained results indicate a high efficacy of pu18 as a potent PS for the use in PDT. Nevertheless, there are still big weaknesses of high pu18 hydrophobicity (leading to the formation of pu18 aggregates), low bioavailability (difficult delivery of pu18 to a tumor tissue), and rapid removal from the tumor tissue [33]. Researchers have tried to solve these shortcomings by derivatization of the pu18 structure, such as creating imides, or anhydrides [34–37], conjugating it to a molecule of a carrier or creating various types of pu18-containing nanoparticles.

A solution to reduce pu18 hydrophobicity and improve its bioavailability is, for example, conjugation of pu18 with a polyethylene glycol (PEG) spacer, as described in our previous article [38]. It was observed that the newly prepared PEGylated pu18 conjugates efficiently accumulated in human cancer cells of various origins (breast—MCF-7, prostate—PC-3 and LNCaP, cervix—HeLa, and pancreas—MiaPaCa-2) in vitro with preferential localization in PDT of important organelles such as mitochondria and endoplasmic reticulum (ER). Attachment of the PEG_3 arm also led to an increase in phototoxicity after light illumination (4 J·cm^{-2}, 13 min) compared to the parental compound. The increase in phototoxicity was probably due to the increase in the conjugate hydrophilicity and, thus, improved solubility of the compound in aqueous media. This further led to a reduction in the formation of pu18 aggregates and better drug bioavailability. To increase the PEGylated conjugate modality, it was further modified with a zinc ion which was incorporated into the pu18 structure. The presence of a zinc ion led to an increase in the absorption in the red region of the visible spectrum, which could enable also imaging of this compound.

Besides PEGylation, conjugating of pu18 to biocompatible graphene oxide (GO), has been described. Thanks to its unique surface properties (carboxyl, hydroxyl, and epoxy groups, free π electron, pH-dependent negative surface charge), GO can bind drug molecules and, thus, contribute to their efficient transport into tumor cells. The properties of GO as a PS carrier were used in the work of Zhang et al. [39], who prepared GO-Pu18 conjugates composed of GO with non-covalently bound (π–π stacking, hydrophobic interactions) pu18 methyl ester, with the aim to increase water solubility, improve bioavailability, control pH-release of pu18 and possibly use in theranostics (see Chapter 9). A similar approach was used by Kang et al. [40], who prepared two types of nanoparticles GO-PS2 and GO-PS3, composed of GO as a carrier and pu18-*N*-ethylamine as a PS. These nanoparticles differed from each other in the type of bonds in the molecule, GO-PS2 contained covalently bound PS, which led to the high stability of the particles under different conditions. In contrast, GO-PS3 contained, like the aforementioned GO-pu18, non-covalently bound PS, which was released at a slightly acidic pH (Figure 4A). Based on the microscopic examination, it was found that all prepared conjugates (GO-Pu18, GO-PS2, GO-PS3) are transported into human tumor cells (HepG2, A549) to a greater extent than the unconjugated pu18. Increased phototoxicity of conjugates GO-Pu18 (HepG2 cell line, xenon lamp, 700 nm, 5 mW·cm^{-2}, 10 min) and GO-PS2, GO-PS3 (A549 cell line, laser, 610–710 nm, 2 J·cm^{-2}, 15 min) vs. free pu18 after photoactivation was detected. Dark toxicity (without photoactivation) corresponded to a free PS for GO-pu18 and GO-PS2 conjugates. No toxicity was observed for GO-PS3 in the specified concentration range. Although GO-PS3 showed a lower rate of $^1O_2$ generation compared to free pu18 (in vitro), higher PDT activity was determined for GO-PS3. This was probably due to the more efficient delivery of GO-PS3 to the cells and the release of non-covalently bound pu18-*N*-ethylamine in the acidic environment of the tumor cells. Both research groups have shown that GO can be a suitable carrier for the efficient delivery of a PS to cancer cells. From the point of view of the targeted release of a PS in tumor tissue, it is advantageous to use noncovalent binding for linking the carrier to the PS molecule.

Currently, great potential in the development of PSs for PDT have nanoparticles. These can be used, for example, for reducing the PS hydrophobicity and to improve its bioavailability. This approach was chosen by Liu et al. [41], who prepared three types of water-soluble organic nanoparticles (WSONs) containing chlorine derivatives such as PS: methyl pyropheophorbide *a* (MPPa), zinc metal-introduced MPPa (ZnMPPa), and pu18 methyl ester (pu18ME). All prepared WSONs reached a size of 87–112 nm and showed good solubility and stability in water. Fluorescence microscopy analysis showed the entry of all WSONs into human HeLa tumor cells and their localization in mitochondria. High phototoxicity of WSONs after light-emitting diode (LED) illumination was also recorded (640–710 nm, 2 $J{\cdot}cm^{-2}$, 15 min) which corresponded to the parental PS. The newly prepared WSONs thus preserve or even improve the properties of the original PS, such as absorption in the red region of the visible spectrum, high phototoxicity, and reduced dark toxicity. Besides, the high solubility of WSONs ensures good bioavailability compared to free PS.

Additionally, other research groups have been involved in increasing PS bioavailability. Since gold nanoparticles (GNPs) represent a promising system for pu18 delivery and ensure better entry into tumor cells, recently the group of Lie et al. [42] utilized this tool. They synthesized ionic liquid (IL)-dependent GNPs with pu18 (Figure 4B). This system is based on the properties of water-soluble ILs that can promote increased solubility of conjugated PSs [42]. It has also been reported that ILs can increase drug stability, promote drug delivery, and controlled release [43]. The prepared drugs consisted of IL-type PSs (PS1 = pu18 + morpholinium, PS2 = pu18 + imidazolium), which were incorporated into GNPs (GNP1 contains 492 $nmol{\cdot}mg^{-1}$ PS1 and GNP2 contains 573 $nmol{\cdot}mg^{-1}$ PS2) to give 20–60 nm large spherical particles. The phototoxicity of the newly prepared GNP1 and GNP2 was verified using the A549 cell line and illumination with a light-emitting diode (735–785 nm, 2 $J{\cdot}cm^{-2}$, 15 min). In the case of GNP1, a 34- and 37-fold increase in phototoxicity was observed compared to free PS1 (12 and 24 h after photoactivation, constant IC_{50} = 0.38 $g{\cdot}mL^{-1}$) and no observable dark toxicity. Besides, microscopic analysis of GNP1-treated cells in combination with light showed the presence of apoptotic bodies. In contrast, GNP2 showed low phototoxicity at 12 h after illumination (IC_{50} > 20 $\mu g{\cdot}mL^{-1}$) and after 24 h, only 3.6-fold higher phototoxicity (IC_{50} = 4.34 $\mu g{\cdot}mL^{-1}$) compared to free PS2. Besides, moderate dark toxicity was determined for both GNP2 and PS2. Thus, it is clear that in the formation of nanoparticles, the type of IL used, and its properties were of great importance.

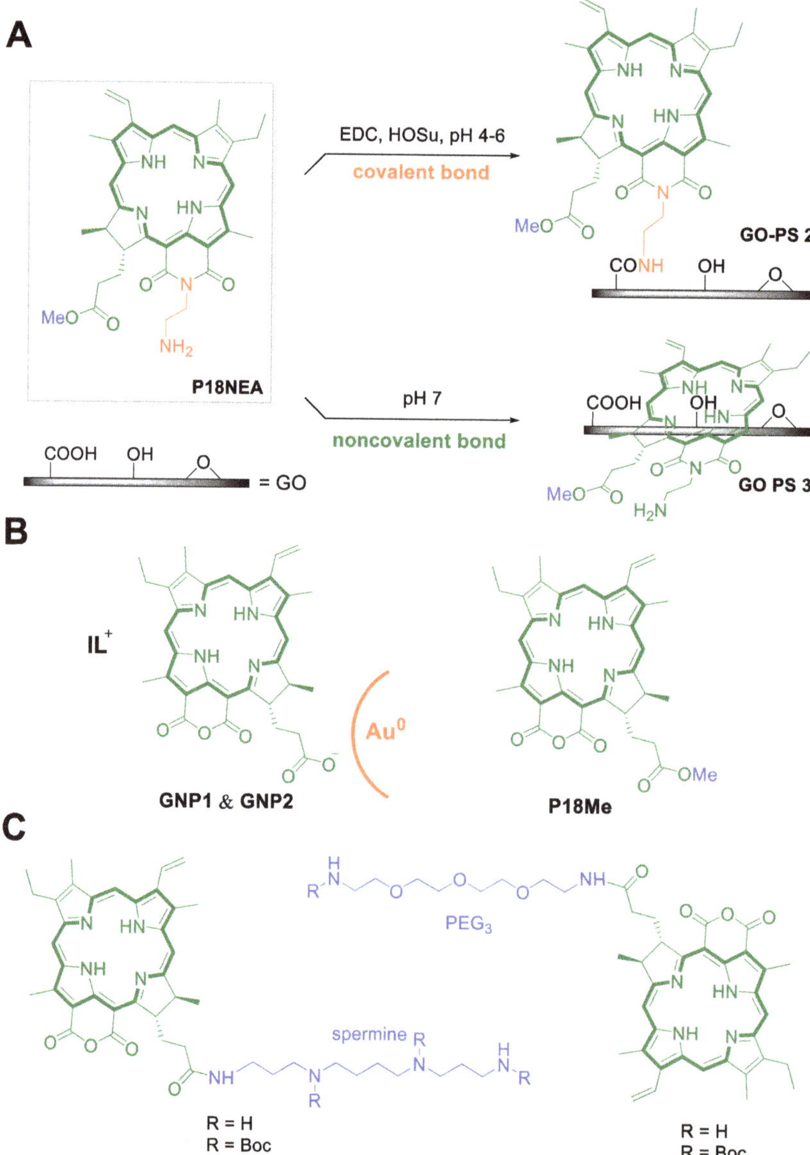

Figure 4. Various types of modifications of purpurin 18 (pu18) may lead to an improvement in its properties as a photosensitizer for use in photodynamic therapy. (**A**) The first approach is to use a suitable carrier such as graphene oxide (GO) to increase the bioavailability of pu18. In this case, two types of GO nanoparticles were prepared with pu18-*N*-ethylamine (pu18NEA) and a photosensitizer (PS), via covalent (GO–PS2) and non-covalent bonds (GO–PS3). (**B**) Another type of nanoparticles is based on the properties of ionic liquids (IL), which together with pu18 were incorporated into gold nanoparticles (GNP1 and GNP2) and the structure of pu18-methyl ester (pu18Me). (**C**) Improved PS retention in tumor cells was observed in the case of conjugated drug pu18 with the amine spermine. In the case of the attachment of a poly(ethylene) glycol (PEG) spacer, an increase in hydrophobicity and an improvement in the bioavailability of PS were observed [44].

Nanoparticles can also play an important role in targeted drug delivery to minimize treatment side effects. As an example, SiO_2 nanoparticles (SiO_2NPs) [45] have several advantages, in particular, nontoxicity and high stability as well as compatibility with various PSs including pu18, chlorine [46], methylene blue [47], and hematoporphyrin [48,49]. The group of Cao et al. used mesenchymal stem cells (MSCs) as pu18 transporters in the form of PS-SiO_2NPs nanoparticles for targeting breast cancer tumor cells [45]. It is known that MSCs naturally show a high affinity for tumor cells [50,51]. The generated PS-SiO_2NPs-MSCs particles migrated to human breast cancer cells (MCF-7), which means that the content of PS-SiO_2NPs does not negatively affect the targeting of MSCs and pu18 can thus be accumulated in targeted cells. At the same time, it was found that PS-SiO_2NPs-MSCs inhibited the growth of MCF-7 cells after PDT treatment, both in the form of 2D culture in vitro and in vivo in nude mice with subcutaneously transplanted MCF-7 cells, in which there was a reduction in tumor size and weight. When PS-SiO_2NPs were introduced into human embryonic kidney cells (HEK-293) instead of MSCs, no significant PDT effect on cell viability was detected since there was no targeted localization of the drug into the tumor cells [38]. The use of MSCs seems a promising strategy for targeted drug delivery to tumor cells.

A long-term approach in the treatment of cancer is a so-called multimodal therapy which may lead mainly to a reduction in the resistance of tumor cells to anticancer chemotherapeutics. This may be, for example, a combination of tumor resection with chemotherapy, a combination of chemotherapy with PDT, or administration of a multimodal drug. Recently, researchers at Southwest University have developed photo-/chemotherapeutics containing pu18 as a PS [52,53]. The first is a nanogel-based drug (DHP NGs ~ 67 nm) composed of pu18, the chemotherapeutic 10-hydroxycamptothecin (HCPT), a redox responsive cross-linking agent, glutathione responsive spacer, and a PEG spacer to increase particle stability. The second multimodal drug contains doxorubicin (DOX) and pu18 encapsulated in PDP micelles (~148 nm) composed of mono (6-amino-6-deoxy)-β-cyclodextrin with a ROS-sensitive polymer containing borate bonds. Both types of drugs have been designed to preferentially accumulate in a tumor tissue based on enhanced permeation and retention effects, for tumor tissue imaging as well as selective drug release due to increased GSH/ROS production by tumor cells. In mouse tumor cell line 4T1, the ability to enter cells, localize in lysosomes and mitochondria, reduce membrane potential of mitochondria and cause cell death has been confirmed [53]. In assays in Balb/c mice implanted with 4T1 cells, longer retention time, and better distribution of DHP NGs in tumor tissue compared to free pu18 were detected, as well as a significant reduction of tumor volume after laser exposure (660 nm, 0.5 $W \cdot cm^{-2}$, 5 min). Immunohistochemical analysis of tumor cells subsequently revealed a large number of necrotic and apoptotic cells due to combined photo-/chemotherapy, the treatment did not cause any obvious damage to other organs (the liver, kidneys, heart, and lungs). Very similar results were obtained for the second drug PDP in combination with the same laser treatment. Upon this treatment, only 17.17% of 4T1 cells survived compared to 40.89 % for pure DOX and 61.28% for pure pu18 in combination with irradiation. Very similar results, as for the first drug, were shown also in experiments on Balb/c mice implanted with 4T1. Both drugs thus represent a significant step in the combined therapy of tumors. A summary of the aforementioned studies on pu18 and its derivatives in PDT is covered in Table 1.

Table 1. Summary of purpurin 18 (pu18) applications and its derivatives aimed for photodynamic therapy (PDT) using human (HL-60, LNCaP, PC-3, MiaPaCa-2, HeLa, MCF-7, HepG2, A549) and mouse (4T1) cell line models. IC_{50}—half-maximal inhibitory concentration after 24 h treatment with pu18, Light—after pu18 photoactivation, dark—without photoactivation, growth inhibition—in studies, where no IC_{50} was calculated, the compound potency was expressed by percentage of growth inhibition at a certain concentration.

Drug	Cell Line	Light Dose	IC_{50} Light	IC_{50} Dark	Growth Inhibition Light	Ref.
Pu18	Human fibroblast	5 W·m^{-2}	-	-	50% (from 10 μM)	[18]
	HL-60	1 J·cm^{-2}	-	-	from 0.2 μM	[31]
	4T1	0.3 J·cm^{-2}	-	>10 μM	>50% (from 0.5 μM)	[33]
		0.74 J·cm^{-2}	-			
		1.59 J·cm^{-2}	-		>50% (from 0.25 μM)	
		3.7 J·cm^{-2}	-			
Pu18+PEG_3 [1]	LNCa	4 J·cm^{-2}	0.02 μM	>10 μM		[38]
	PPC-3		0.65 μM			
	U-2 OS		1.83 μM			
	MiaPaCa-2		0.45 μM			
	MCF-7		0.59 μM			
	HeLa		0.02			
GO[6]-Pu18	HepG2	5 mW·cm^{-2}	1.4 mg·mL^{-1}	>5 mg·mL^{-1}		[39]
GO[6]-PS2	A459	2 J·cm^{-2}	0.31 μM	>14 μM		[40]
GO[6]-PS3			0.20 μM			
Pu18ME-WSON [2]	HeLa		0.83 μM	>10 μM		[41]
GNP1 [3]	A549		0.38 μg·mL^{-1}	>20 μg·mL^{-1}		[42]
GNP2 [3]			4.34 μg·mL^{-1}			
DPH NGs [4]	4T1	0.5 W·cm^{-2}	-	-	>50% (HCPT [8]: 2 μg·mL^{-1}, pu18: 5 μg·mL^{-1})	[53]
Pu18			-	-	38.72%(8.7 μg·mL^{-1})	[51]
PDP [5]			-	-	82.83% (DOX [7]: 5 μg·mL^{-1}, p18: 8.7 μg·mL^{-1})	

[1] PEG—poly(ethylene) glycol; [2] WSON—water-soluble organic nanoparticles; [3] GNP—gold nanoparticles; [4] DPH NGS—designed prodrug nanogels; [5] PDP—cyclodextrin-based micelles; [6] GO—graphene oxide; [7] DOX—doxorubicin; [8] HCPT—10-Hydroxycamptothecin.

5. Sonodynamic Therapy and Sonophotodynamic Therapy

Another noninvasive approach of cancer treatment is sonodynamic therapy (SDT). It is a method discovered and derived based on PDT. In PS evaluation, it was found that, for example, porphyrin [54–57], pheophorbide [57,58], and/or pu18 derivatives [59] used in PDT show similar effects and cell damage after ultrasound application as detected after photoactivation. The main principle of SDT is identical to PDT. The SDT method is based on a combination of low-intensity ultrasound and a sonosensitive agent (ST), which is activated by ultrasound (Figure 5). However, the mechanism of ST activation and ROS formation in SDT is not exactly known. The action of ultrasound generates heat in the target tissue and, also, acoustic cavities. These can occur in two states: stable cavities, which oscillate in an aqueous environment, leading to fluid flow, and inertial cavities, which are characterized by rapid enlargement and subsequent sharp collapse (implosion). During the collapse of these cavities, a large amount of heat (≥5000 K) and pressure (≥800 atm.) is released [60,61]. This can lead to a series of chemical reactions inside, on, or around the collapsing cavitation bubble, which is often referred to as a sonochemical reactor [62]. Typically, pyrolysis or thermal dissociation of water may occur here, which may lead to ST activation and, also, to ROS production.

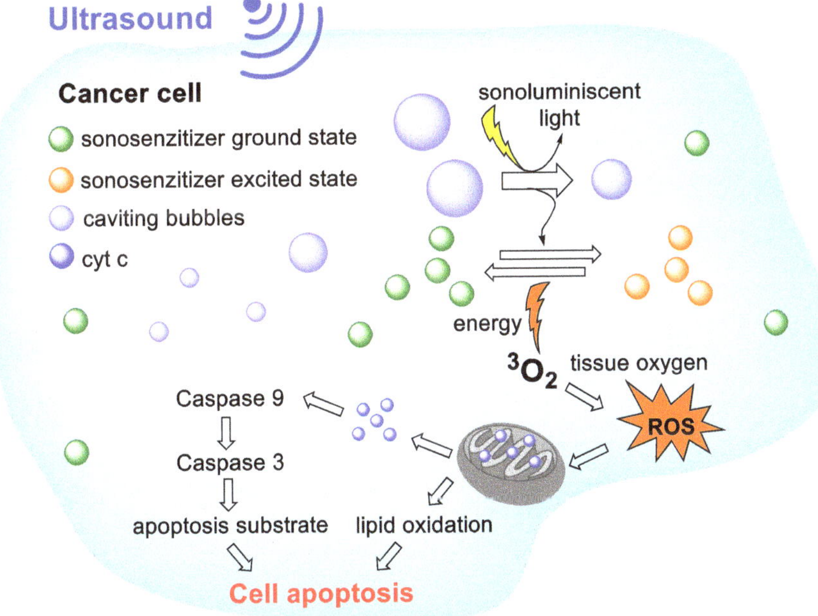

Figure 5. General principles of sonodynamic therapy. The action of ultrasound activates the sonosensitive agent and its transition to a higher energy state. When returning to the basic energy state, the energy generated is transformed in the form of heat, acoustic cavities, or light (sonoluminescence). As a result of these processes, reactive oxygen species (ROS) are formed, which leads to damage of the mitochondrial membrane, the release of cytochrome c (cyt c), and triggering a cascade leading to cell death.

However, some authors suggest that ST activation occurs differently due to sonoluminescence. Sonoluminescence is described as the emission of light that can occur when cavities collapse. This emitted light could thus activate the ST, similarly to that in PDT. However, scientists do not fully agree on this mechanism either [63]. It is, thus, likely that the mechanism of SDT action is a combination of several factors, which depend on the ST and the parameters of the ultrasound used. Although the exact mode of action of SDT has not been elucidated, SDT has tremendous potential in the treatment of cancers. An indisputable advantage of SDT over PDT is the deep penetration of ultrasound into soft tissues, which can be even more than 10 cm [64]. As some PSs can be used as STs, it is possible to combine both approaches as SDT-PDT.

Recently, Shi et al. [65] have described the development of a new tumor-specific and multi-activatable drug called nano-riceball that can be used in SDT. This is a multilayer micelle with a size of 96.8 nm, the core of which is constructed of amphipathic bottlebrush-like dextran-based polymer (DOS), which carries in its structure the anticancer drug DOX and pu18, as a potent ST. This construct is denoted as DDP. For specific tumor cell targeting, the upper layer of the DDP is covered by a bioengineered cell membrane [66], which carries the target motif of the Asn-Gly-Arg (NGR) peptide [67]. It is recognized by various isoforms of the CD13 receptor (aminopeptidase N), which occurs on the membrane of many types of tumor cells and, thus, it is a suitable target for the delivery of anticancer drugs (Figure 6). In this way-targeted drug nano-riceball (NGR® DDP) is preferentially captured on the surface of tumor cells, and is transported into the cells by endocytosis. Shi et al. [68] found that NGR®DDP accumulates in HepG2 cells within 3 h of treatment. After sonication of the cells ($0.56\ W \cdot cm^{-2}$, 5 min), there is a rapid increase in ROS, lysosome disintegration, and collapse of the NGR@DDP structure. This leads to a partial release of DOX, then, additional DOX release is achieved by endogenous H_2O_2 stimulation. By application of ultrasound, pu18 is also activated, which leads to further ROS generation. Due to the fluorescence of pu18, drug localization can be also monitored. Tests in nude mice with Luc-HepG2 tumors showed preferential accumulation of NGR® DDP in tumor tissue as well as significant reduction or complete disappearance of the tumor after administration of NGR® DDP in combination with an ultrasound. Besides, histological analysis showed extensive tumor tissue damage with minimal damage of normal/healthy tissues, thus suggesting high drug selectivity with the generation of minimal side effects.

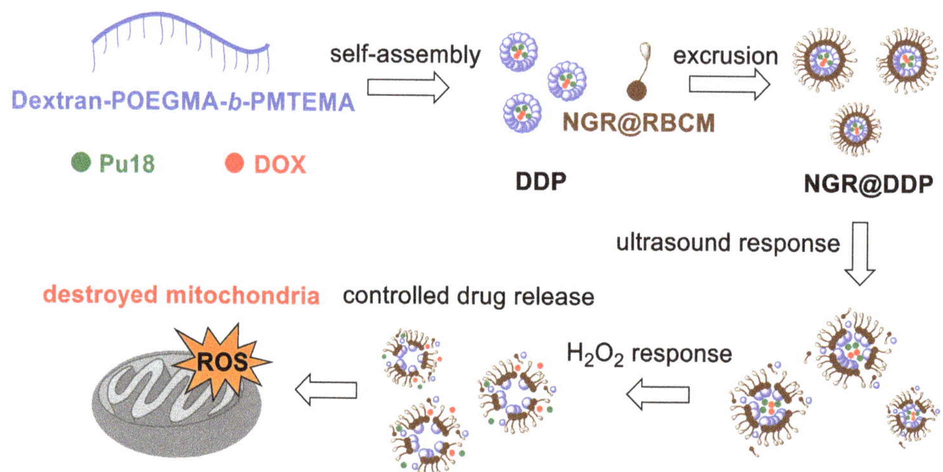

Figure 6. Schematic representation of the preparation of the tumor-specific, mitochondrially targeted, multi-activated (ultrasound/H_2O_2) drug NGR®DDP. These are micelles consisting of a universal amphipathic bottlebrush-like polymer dextran-POEGMA-b-PMTEMA, the chemotherapeutic doxorubicin (DOX), and purpurin 18 (pu18), collectively referred to as DDP. These micelles are further coated with a top layer (NGR®RBCM) which carries the target motif of the Asn-Gly-Arg (NGR) peptide. POEGMA = poly(oligo ethylene glycol methacrylate); PMTEMA = poly [2-(methylthio)ethyl methacrylate]; NGR®RBCM = peptide-labeled red blood cell membrane.

The ST drug PTPK has been developed to image and suppress the growth of pancreatic tumors [68]. Hydrophilic PTPK was prepared in the form of a linear chain composed of the cytotoxic peptide Lys-Lys-Leu-Ala-Lys-Leu-Ala-Lys-Lys-Leu-Ala-Lys-Leu-Ala-Lys (KLAK-LAK), the ST pu18, and PEG, which is attached by a thiol bond. After PTPK treatment with ultrasound, pu18 is activated to form singlet oxygen, which leads to cleavage of the thiol bond and cleavage of the PEG spacer. The rest of the SH-(pu18) KLAK molecule then forms self-assembled nanoparticles due to increased hydrophobicity. The nanoparticles are then transported into cells by endocytosis (Figure 7). Cytotoxicity assays on the Panc-1 cell line show that PTPK exhibits higher cytotoxicity and higher mitochondrial damage to treated cells after ultrasound treatment compared to non-cleavable PEG (pu18) KLAK and pre-prepared (pu18)-KLAK nanoparticles (without PEG). Using a multicellular tumor spheroid of Panc-1 cancer cells, the ability of single-stranded PTPKs to penetrate deeper into tissues was observed than with pre-packaged (pu18)-KLAK nanoparticles. In situ packaging of nanoparticles is also likely to lead to improved cell transport through endocytosis. Tests on nude mice implanted with Panc-1 cells have shown that PTPK in combination with ultrasound inhibits the growth of subcutaneous and orthotopic tumors.

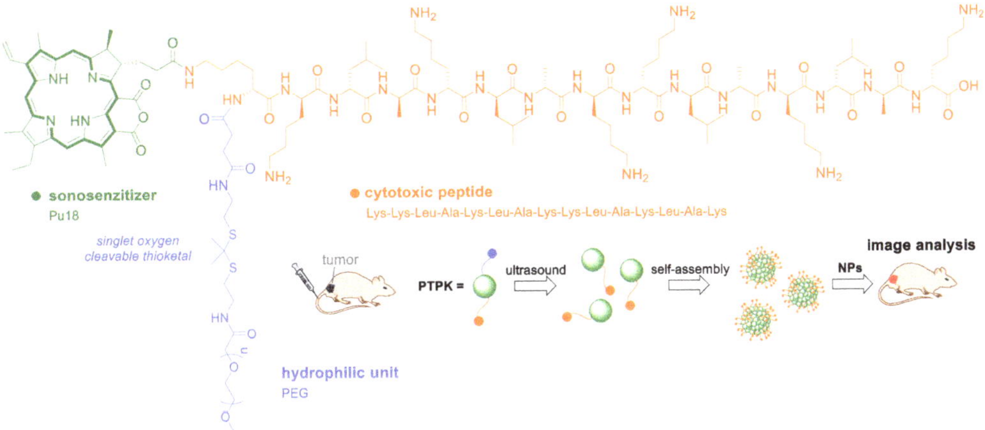

Figure 7. Schematic representation of the ultrasound activatable polypeptide drug PTPK. The drug is applied to the bloodstream in a linear form, pu18 is activated to form singlet oxygen, which leads to cleavage of the thiol bond and disconnection of the PEG spacer. The rest of the molecule then consists of hydrophobic nanoparticles, which can more easily penetrate tumor cells.

6. Photoacoustic Imaging and Therapy

Photoacoustic imaging (PAI) is a promising technique for diagnostic and therapeutic purposes in the treatment of cancer due to its high resolution and deep penetration into tissues [69]. In medical applications, photoacoustic imaging is currently in clinical trials for imaging tumors of the breast [70,71], head and neck [72,73], thyroid gland [74], colon, and rectum [75]. The principle of PAI is the generation of acoustic ultrasonic waves based on the conversion of energy supplied by pulsed laser light. If a suitable molecule absorbs light energy, this energy is converted into heat, which also leads to an increase in the temperature around the molecule. Based on the thermoelastic expansion caused by the increase in temperature, ultrasonic waves are subsequently generated, recorded, and analyzed to form an image. Naturally occurring chromophores such as oxy-/deoxyhemoglobin, melanin, or exogenous inorganic and organic substances, which have the potential to absorb energy in the near-infrared region of the spectrum, are used as contrast agents in PAI. Endogenous chromophores can be used in a variety of applications, from brain mapping to monitoring angiogenesis and hypoxia, which are one of the main features of tumor presence. However, their main advantage, i.e., endogenous occurrence, is also their main limitation, due to the difficult distinction between the increased signal vs. the background signal (e.g., in tumor imaging). Some PSs, such as indocyanine green [76], and methylene blue [77] have been approved by the US Food and Drug Administration for application in PAI. Besides, due to absorption in long wavelengths, pu18 and its derivatives are also suitable for such application.

For example, it could be beneficial in distinguishing very small tumor features (<10 mm) from healthy tissues. PAI probes such as pu18/AHC⊂L (AHC⊂L—ammonium hydrogen carbonate encapsulated in liposomes) and CD44v6-pu18/AHC⊂L with improved photoacoustic properties are nanoparticles composed of a lipid bilayer, in which pu18 and NH_4HCO_3 aggregates are incorporated [78]. NH_4HCO_3 is in the hydrophilic core of the particle (pu18/AHC⊂L). After laser irradiation of the particles, pu18 aggregates are activated to produce heat and photoacoustic signals (Figure 8). The generated heat (42 °C) further initiates the slow decomposition of NH_4HCO_3, which leads to the formation of CO_2 bubbles, which significantly contribute to the long-term increase in the photoacoustic signal (8 h). For targeted imaging of bladder tumors, the surface of the nanoparticles was conjugated to an antibody against the surface protein CD44v6 (CD44v6-pu18/AHC⊂L). This protein has been identified in the colon [79,80], breast [81], and bladder [82] cancer cells, where it is produced to a greater extent than in healthy tissues. The CD44v6 protein is associated with cell migration, adhesion, and tumor metastasis. By comparing the effect of CD44v6-pu18/AHC⊂L on tumor and healthy human bladder tissue, a preferential accumulation of CD44v6-pu18/AHC⊂L was found in tumor tissue with a clear tumor delineation. Very small lesions (<5 mm) were also distinguished, which is a great advantage for early diagnosis of the tumor.

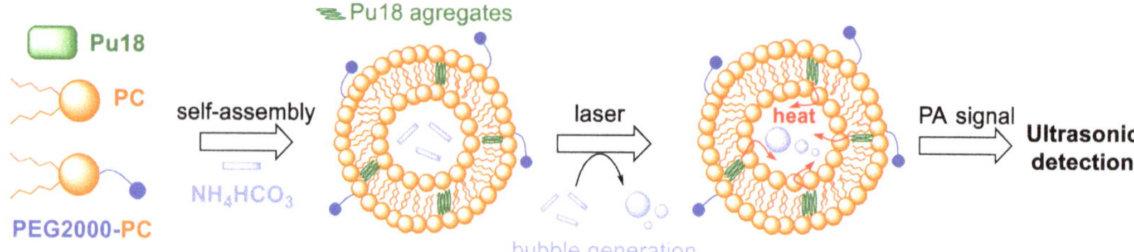

Figure 8. The diagram shows the structure of nanoparticles composed of a lipid bilayer (L-α-Phosphatidylcholine = PC, poly(ethylene glycol) 2000 = PEG2000), in which aggregates of purpurin 18 (pu18) and NH_4HCO_3 (in the hydrophilic core of the particle) are incorporated. Under the action of laser irradiation, the pu18 aggregates are activated to produce heat and a photoacoustic (PA) signal. The generated heat activates the slow decomposition of NH_4HCO_3, with the formation of CO_2 bubbles. This ensures a long-term increase in the PA signal, which can be easily detected for several hours [78].

The problem of the application of not only PAI drugs is bioavailability, which significantly affects the success of the whole process. An attractive approach of the preparation of new bioavailable drugs is the creation of so-called self-assembly molecules. These are usually multifunctional nanostructures that are formed based on van der Waals or electrostatic forces, hydrogen bonds, hydrophobic or π-π interactions directly in biological systems. This approach was used by Zhang et al. [26] to prepare self-assembled PAI nanofibers that are formed by precursor units. The precursor (pu18-PLGVRGRGD) is composed of pu18 as a PA functional molecule, the enzymatically cleavable peptide chain Pro-Leu-Gly-Val-Arg-Gly (PLGVRG), and the Arg-Gly-Asp (RGD) sequence targeting $\alpha v\beta 3$ integrins as tumor cell surface structures. Under the action of the enzyme gelatinase, which occurs in tumor cells, the structure of pu18-PLGVRGRGD is cleaved. The remaining pu18-PLG units subsequently form nanofibers due to intramolecular π-π interactions (Figure 9). Comparison of the effect of pu18-PLGVRGRGD with a pu18-PLGVRGRDG molecule (without target sequence) on $\alpha v\beta 3$ overexpressing U87 cells integrates and gelatinizes, confirming the crucial role of the target motif in PA probe accumulation in tumor cells. In mice with implanted U87 cells, the PA signal increased already 2 h after pu18-PLGVRGRGD application. The highest PA signal was monitored 6 h after application and a stable PA signal could be observed for the next 24 h. Observed PA signal after application of pu18-PLGVRGRGD, range of double values compared to pu18-PLGVRGRDG (no target sequence), indicating increased accumulation of pu18-PLGVRGRGD in the tumor. The formation of nanofibers directly at the tumor site can also lead to a prolongation of the retention time of the probe in the tumor tissue (so-called assembly/aggregation-induced retention). The half-life of pu18-PLGVRGRGD is almost 24 h, which is six times more than in the case of the control probe pu18-PMGMRGRGD with a change in the peptide sequence that is not recognized by gelatinase (does not form nanofibers). In a long-term experiment with U87-xenografted mouse cells, complete tumor disappearance occurred after 23 days of treatment with the pu18-PLGVRGRGD probe in combination with laser treatment. The histopathological analysis did not prove damage to other organs (the heart, liver, kidneys, lungs), while the tumor tissue showed signs of necrosis.

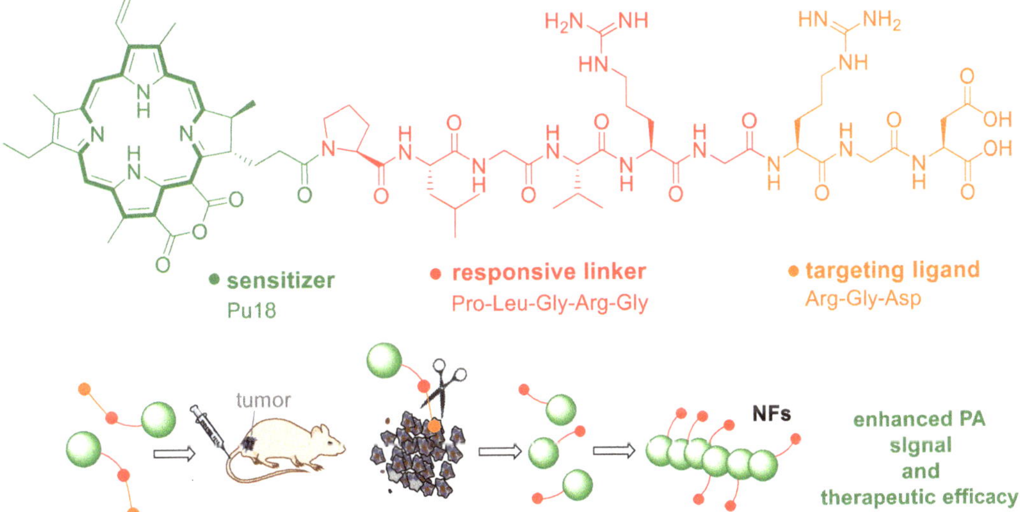

Figure 9. The scheme shows the structure of a prodrug (pu18-PLGVRGRGD) for the preparation of a targeted photoacoustic drug. The pu18-PLGVRGRGD precursor consists of purpurin 18 (pu18), Pro-Leu-Gly-Val-Arg-Gly (PLGVRG) as the enzyme-responsive peptide linker, and Arg-Gly-Asp (RGD) as the targeting ligand. Following application of the monomer precursor, it cleaves at the PLGVRG site in tumor cells. The remaining pu18-PLG units form nanofibers (NFs), which exhibit an enhanced photoacoustic signal [26].

The self-assembly of nanofibers from preformed nanoparticles directly inside cells was an approach used by Zhang et al. [83]. This study was based on the assumption that the formation of nanofibers is dependent on several factors such as concentration of the precursor, ionic strength, and temperature. A transformable polymer-peptide conjugate PKK-S-PEGm was created, which were composed of the cytotoxic peptide KLAKLAK that binds to the mitochondrial membrane, hydrogen-bonding peptide Lys-Leu-Val-Phe-Phe (KLVFF), glutathione-responsive motif disulfide bond linked to PEG chain, and photoacoustic/photothermal molecule pu18 [83]. PKK-S-PEG probes were prepared in the form of 40-nm nanoparticles, which are transported into cells by endocytosis. Upon entry inside the cells, disulfide bridges are cleaved by glutathione, thereby removing the PEG chain stabilizing the spherical structure of the nanoparticle. The nanoparticles are then gradually transformed into nanofibers, which are formed owing to the hydrogen bonds of the KLVFF motif. The morphological change of nanoparticles into nanofibers can be accelerated by laser photoactivation of the particles. Due to the presence of pu18, heat is generated after laser illumination, which accelerates the transformation of nanoparticles into nanofibers up to four times (Figure 10). Cytotoxicity assays on the HeLa, MCF-7, and L929 cell lines showed increased cytotoxicity and higher number of apoptotic cells treated with PKK-S-PEG in combination with laser illumination than in non-illuminated cells. Due to the assay on BALB/c mice transplanted with Hela cells, increased accumulation and retention time of the drug in the tumor tissue was observed. Due to the formation of photoacoustic waves, it was possible to monitor the formation of nanofibers and the movement of the drug in real-time.

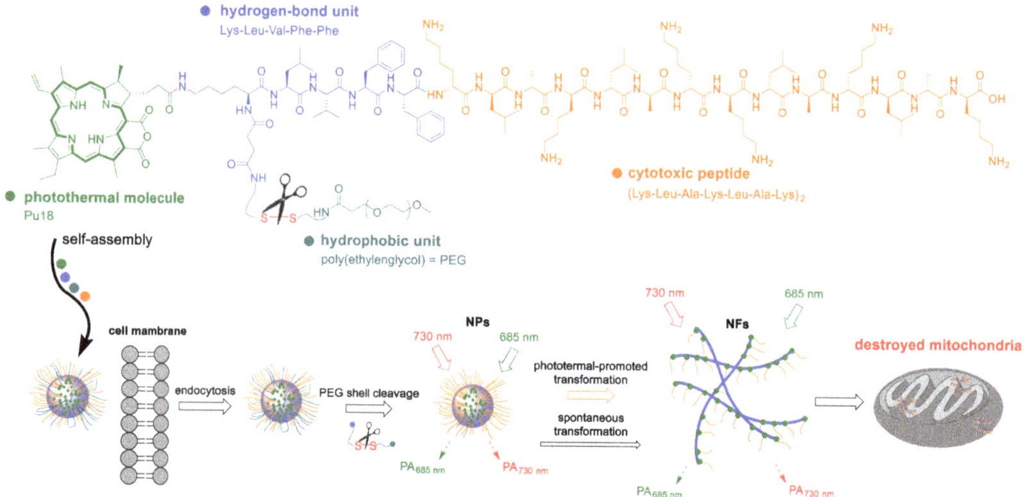

Figure 10. The diagram shows the chemical structure of the polymer-peptide conjugate PKK-S-PEG. It is composed of the cytotoxic peptide Lys-Leu-Ala-Lys-Lu-Ala-Lys (KLAKLAK), which binds to the mitochondrial membrane, a hydrogen-bonding peptide motif Lys-Leu-Val-Phe-Phe (KLVFF, originated from the Aβ protein), the hydrophobic unit of poly(ethylene glycol) = PEG, which is attached via a disulfide bridge, and the photoacoustic/photothermal molecule pu18. Drug monomers are spontaneously packaged into nanoparticles, which are transported into cells by endocytosis. Upon entering the cells, the PEG chain is cleaved leading to the spontaneous transformation of nanoparticles (NPs) into mitochondrially targeted nanofibers (NFs). The conversion in NFs can be accelerated photothermally [83].

A similar approach was used by Zhou et al. [25], who studied the properties of self-assembly of molecules composed of pu18 and the amino acid arginine to monitor the aggregation process by PAI in vivo [25]. The prepared derivatives PA1, PA3, and PA7 differed from each other in the number of arginine units (1, 3, 7), from which the spectral properties, the ability to form aggregates, and the morphology of the formed particles are derived. In the case of monomeric pu18, there was a redshift of the Q_y band from 700 to 750 nm during the aggregation process, which indicates the formation of J-type aggregates. At the same time, the fluorescence signal gradually decreased. In the case of monomers PA1, PA3, and PA7, the extinction coefficients (at 700 nm, in dimethylsulfoxide) increased at the same concentration together with the increasing number of arginine units. In an aqueous environment, where conjugates aggregated, higher solubility was observed with a higher number of arginine units leading to less redshift due to hydrophobicity and positive charge of the conjugates. With a higher number of arginine units, a reduced ability to form aggregates and a weak π-π interaction were also observed for PA7. The same result was shown by the measurement of the fluorescence intensity; in the case of PA1, there was a complete loss of fluorescence, which indicates that all monomers undergo to an aggregated state. In contrast, PA7 showed a similar level of fluorescence in the aqueous medium as in the monomeric state. The fluorescence intensity of PA3 was reduced in the aqueous medium, but it was not completely lost. In aqueous media, both PA3 and PA7 are likely to occur in both monomeric and aggregated states. On the contrary, it is known that as the fluorescence signal decreases, the PA signal increases. In connection with the photoacoustic signal, the heat conversion efficiency of the conjugates was studied. A 3-fold increase in heat conversion efficiency was observed for the PA1 and PA3 aggregates compared to their monomeric state. An increase in heat conversion efficiency was also observed for PA7, although not to such an extent. The degree of aggregation of the conjugates in the aqueous medium correlated with heat conversion efficiency, which is proportional to the photoacoustic signal. Zhou et al. [25] also defined the relationship between the degree of aggregation and the relative photoacoustic ratio as a useful tool for aggregation research.

RGD-Dex/NPBA-pu18 (RDNP) monomers were generated to monitor targeted drug delivery, and the structural changes induced by the acidic environment of the cells. By different molar ratios of biocompatible peptide Cys-Gly-Gly-Arg-Gly-Asp (RGD) with dextran (RGD-Dex) as a carrier and a photoacoustic probe pu18 modified with boric acid (NPBA-pu18), a total of five types of monomers (P1-P5) with different properties were prepared. It was found that when NPBA-pu18 is in a 3-fold and higher molar excess over RGD-Dex, RDNP monomers in an aqueous medium can be self-assembled into nanoparticles [84]. The hydrophobic NPBA-pu18 forms a core of nanoparticles that is surrounded by a hydrophilic RGD-Dex. It serves to stabilize the nanoparticle. In an acidic environment, RDNP may cleave due to a pH-dependent phenyl borate bond. The released NPBA-pu18 can repackage into the structure of highly organized nanofibers due to high hydrophobicity and π-π interactions (Figure 11). After laser irradiation, it is then possible to monitor the PA signal and distinguish the structural changes that are characteristic for the monomeric and aggregated state of NPBA-pu18. P2 (single-stranded) and P5 (nanogranulated) labeled with Cy5 were used for cell entry assays in HeLa cells. Based on the intensity of the fluorescent signal, it was determined that P5 internalized into cells more efficiently than P2 by caveolae-mediated and clathrin-mediated endocytosis. Both conjugates were localized in cell lysosomes/endosomes. 1h after the treatment of HeLa cells with P5, an increasing photoacoustic signal was detected and after 3 h when all nanoparticles were likely to dissociate to form NPBA-pu18 aggregates, a stable PA signal was achieved. The entire process of structural transformation of RDNP was monitored in real-time using PAI, which makes this method an important tool for following many cell processes, such as targeted drug delivery or autophagy, which is often connected to cancer.

In general, autophagy is a natural catabolic process, in which proteins, organelles, and other cell components are degraded. It is one of the essential cellular processes for maintaining homeostasis, which occurs in response to stressful conditions such as hypoxia, presence of ROS, or lack of nutrients. Some drugs, such as DOX, have been reported to up-regulate autophagy, which can, however, paradoxically lead also to resistance of the cell to chemotherapy [85,86]. Therefore, autophagy inhibitors are commonly co-administered with chemotherapeutics to help increase the success of anticancer therapy [87]. However, the combination of these drugs often leads to high systemic toxicity and damage to the organism. Thus, preventive monitoring of autophagy in real time, directly in tumor cells using PAI, could be beneficial. With this goal, Lin et al. [88] developed a novel photoacoustic in vivo self-assembled probe (DDP18, Figure 12). It is composed of hydrophobic pu18 as a signaling molecule, Gly-Lys-Gly-Ser-Phe-Gly-Phe-Thr-Gly (GKGSFGFTG) peptide, the cleavage site of autophagy-related cysteine protease, and poly(amidoamine) dendrimer, which serves as a carrier increasing the solubility and biocompatibility of the molecule. Upon entry of the probe into cells undergoing autophagy, the pu18 residue is enzymatically cleaved with a linker to aid in its folding.

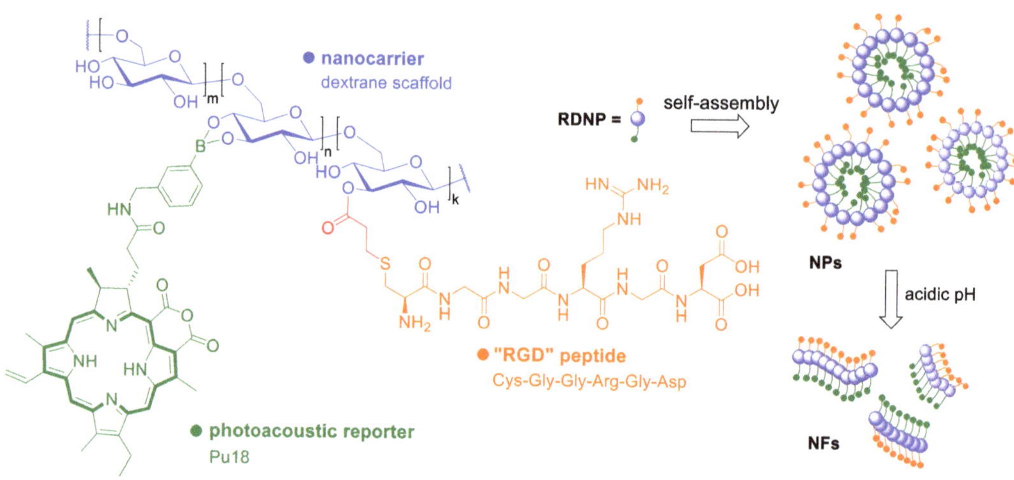

Figure 11. The diagram shows the structure of the photoacoustic drug RDNP. The drug monomers consist of dextran that serves as a carrier, a biocompatible peptide Cys-Gly-Gly-Arg-Gly-Asp (RGD), and a photoacoustic probe pu18, which is modified with boric acid. RDNP monomers spontaneously form nanoparticles (NPs) in an aqueous environment, which can be cleaved to form highly organized nanofibers (NFs) upon transition to the acidic environment of tumor cells. The conversion of NPs to NFs can be monitored in real-time using photoacoustic imaging [84].

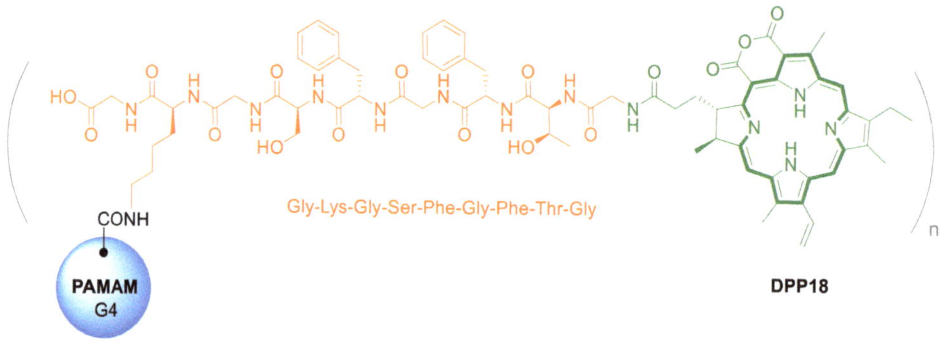

Figure 12. Chemical structure of photoacoustic in vivo self-assembled nanoprobes for imaging and quantitative detection of autophagy. The molecule is composed of poly (amidoamine) dendrimers (PAMAM), the specifically cleavable peptide Gly-Lys-Gly-Ser-Phe-Gly-Phe-Thr-Gly (GKGSFGFTG), and purpurin 18 (pu18) [88].

It has been reported that pu18 in the form of aggregates/nanostructures has a higher photoacoustic response after laser activation than free pu18 due to a change in photothermal conversion capacity [78,89]. Simultaneously with the increase in the PA signal, a decrease in pu18 fluorescence was observed. When verifying the cytotoxic effects in the MCF-7 cell line, very low cytotoxicity was found both without and after laser treatment of the cells under conditions corresponding to PAI (735 nm, 100 mW, 10 s), the rate of the photoacoustic signal was also very low (autophagy was not detected). When cells treated with DPP18 were simultaneously treated with an autophagy inducer, a significant increase in photoacoustic signal was measured after laser irradiation. Using DPP18 in vivo in mice, it has been shown that monitoring tumor autophagy with PAI can be successfully used to optimize the dosing schedule of the chemotherapeutic DOX in combination with the autophagy inhibitor hydroxychloroquine. Mice treated with the adjusted dosing regimen showed similar tumor growth suppression as mice treated with the traditional dosing regimen. Besides, however, no significant weight loss of the mice or a decrease in white blood cell and platelet counts were observed with the adjusted dose, suggesting an overall better physical condition of the animals. The use of PAI can thus be a promising tool for personalized medicine.

7. Photothermal Therapy

The basic component of PTT is, as in the case of PDT, a photosensitive drug molecule that is activated by laser light, usually at near-infrared wavelengths. PTT, however, converts light energy absorbed by a photosensitive drug into thermal energy, which can heat the target tissue to more than 43 °C, leading to DNA damage, denaturation of proteins, and subsequent apoptosis or necrosis. The type of cell death depends on the drug and the type of laser irradiation (laser energy, number of pulses, duration of irradiation) used. As PTT does not require the presence of oxygen in the tissues, it is also suitable for hypoxic tumors, which are often resistant to classical therapy [90]. Nanoparticles of inorganic origin are often used as PTT drugs, including porphyrins, which are usually modified into a nanoparticle structure. However, due to the similarity of PTT and PDT, it is possible to often come across combined drugs that utilize properties of both PTT and PDT to increase the antitumor efficacy. Kim et al. [91] prepared a complex of combined PTT/PDT gold-nanorods (GNR-PS). The formulation is based on GNR composed of hydrogen tetrachloroaurate (III) trihydrate ($HAuCl_4 \cdot 3H_2O$) and cetyltrimethylammonium bromide, which is, however, cytotoxic. For this reason, the layer-by-layer method was used to prepare GNR-PS using cationic poly (allylamine hydrochloride), which suppresses the cytotoxicity of cetyltrimethylammonium bromide and is important for the formation of a dual PTT/PDT system using pu18 as a PS. Verification of the PTT and PDT effects of the GNR-PS drug was performed in A549 cells. After the cell treatment with the GNR-PS, they were photoactivated with a laser (808 nm, 120 $J \cdot cm^{-2}$, 1 min) for PTT induction followed by a halogen PDT-inducing lamp (a band-pass filter 640–710 nm, 2 $J \cdot cm^{-2}$, 15 min). PTT/PDT-treated cells showed the highest decrease in viability (39% of viable cells) compared to single therapies (PDT = 90%, PTT = 73% of viable cells). At the same time, no toxicity of GNR-PS without illumination was observed in the evaluated concentration range (1–20 μM). The results also showed that within 1 min after the PTT, 57% of the pu18 contained in the GNR-PS was released and without photoactivation only 2% within 60 min. The combined GNR-PS drug effectively utilizes the synergistic effect of PTT/PDT and is a potent antitumor drug.

8. Phototeranostic

The aforementioned methods might all be combined. Moreover, it is possible to combine them with diagnostic tools, which is clinically a very interesting and more and more emerging approach, generally called theranostics. Theranostics represents therapy and diagnosis of a disease covered by a single substance or particle, which, in general, can exhibit very diverse biological activities. The therapeutical part may consist of all possible currently known chemotherapeutics, the structure of which can be modified and linked to the diagnostic part. Diagnosis, in this case, is provided usually by radioactivity or fluorescence of the particular substance. Nevertheless, also other clinical diagnostic tools are being involved, such as magnetic resonance, computer tomography, and others [92]. Pu18 and related derivatives are also being developed for such applications. The advantage of pu18 is the ability to absorb light and emit fluorescence. Once delivered into pathological cells, this ability can be utilized for tumor imaging. Zhang et al. 2018 synthesized pH-dependent pu18 loaded GO particles, which were stable at physiological pH, and at low pH, often occurring in tumors, released free pu18. This might be detected by increased fluorescence and, thus, display the tumor. Such particles have also a better ability to yield singlet oxygen (62.6% quantum yield) compared to free pu18 (58.82%). In HepG2 cells, IC_{50} of GO-pu18 after 10 min of irradiation was 1.39 ± 0.38 $\mu g \cdot mL^{-1}$, whereas that of free pu18 was 5.3 ± 0.79 $\mu g \cdot mL^{-1}$. The viability of cells without irradiation was higher than 90% for all the studied concentrations [39]. Generally, pu18 has cytotoxic and fluorescent properties, which makes this compound alone a theranostics agent. Thus, the only question is, how to efficiently and specifically deliver this substance to cancer cells and how to preserve its fluorescent and phototoxic properties. The answer might be again proper derivatization. Sun et al. [93] developed a theranostic system consisting of self-assembled nanoparticles (ppu18-lipos) of lipid-pu18 of 2 mol% and 65 mol% concentrations and pure lipids. The lipid-pu18 derivatives were created by the reaction of pu18 and 1-palmitoyl-2-hydroxy-*sn*-glycero-3-phosphocholine. Ppu18-lipos enabled trimodal in vivo imaging by fluorescent and photoacoustic methods and magnetic resonance (Figure 13). The fluorescent signal was observed in vitro more after low-density particle treatment because pu18 in higher concentrations formed aggregates, which caused fluorescence quenching. Photoacoustic signals, on the contrary, increase linearly with pu18-lipid concentrations, since such signals correlate with the thermal expansion of laser irradiated materials. Magnetic resonance signal, which was enabled by treatment with ppu18-lipos-Mn^{2+} chelates, also increased with ppu18 concentration. A potent therapeutic effect of the ppu18-lipos particles was achieved by PDT and PTT in combination and, also, activation of the immune response was detected after laser irradiation. The synergism of PDT and PTT was followed in vivo in mice bearing 4T1 breast tumors. Astonishing 90% inhibition of tumor growth was achieved with a concentration of ppu18, which was 6 mg·kg^{-1} lower than that for PDT and PTT alone. Importantly, the advantageous layout of the experiment seems to be a pretreatment of low-density particles and laser activation, because PDT ensures more facile endocytosis of further pu18-lipid particles and subsequently more efficient PTT [93].

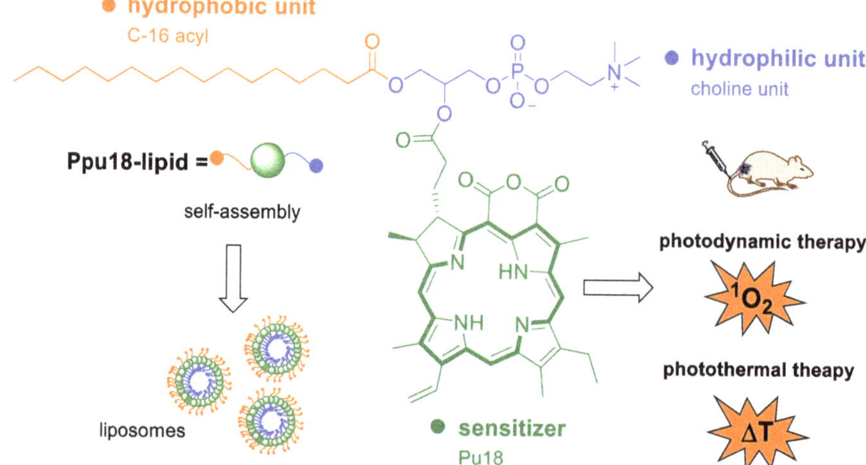

Figure 13. The diagram shows the structure of a combined photodynamic/photothermal drug, which is composed of purpurin 18 (pu18), a hydrophobic part formed by C-16 acyl, and a hydrophilic choline unit. Ppu18-lipid drug monomers form self-assembled liposomes, which can be used as combined theranostics for PDT/fluorescence imaging or PTT/photoacoustic imaging, depending on the amount of pu18 used (mol%) [93].

9. Conclusions

Chlorophyll and its natural metabolites are a very valuable tool in the development of novel anticancer drugs. In particular, one of its metabolites pu18 is a very potent multifunctional molecule that can be used in several anticancer therapies such as PDT, SDT, and PTT, but also in theranostics and tumor imaging for their early diagnosis, e.g., by PAI. All these applications are possible due to the structure of pu18, which allows strong energy absorption and further transfers of this energy leading to the creation of cytotoxic ROS, heat, or fluorescent signal. Another great advantage of pu18 is its low systemic toxicity in the dark environment, thus, causing low side effects during the therapy. Unfortunately, further advancement into clinics is hampered by its low water solubility. However, this problem can be easily overcome by its derivatization, which leads to the synthesis of many potent conjugates. Owing to its facile derivatization and unique properties, pu18 represents an effective tool for the development of novel PSs and multimodal drugs, which seem to be the future of anticancer therapy.

Funding: This research was funded by Specific university research MSMT, grant No. A1_FPBT_2020_001.

Institutional Review Board Statement: Not applicable.

Informed Consent Statement: Not applicable.

Data Availability Statement: Not applicable.

Acknowledgments: The authors would like to thank the *Applied Sciences* journal for providing the APC waiver.

Conflicts of Interest: The authors declare no conflict of interest.

Abbreviations

2D	two-dimensional materials
3D	three-dimensional materials
AHC⊂L	ammonium hydrogen carbonate encapsulated in liposomes
Chl	chlorophyll
Chl*a*	chlorophyll *a*
Chl*b*	chlorophyll *b*
Cyt c	cytochrome c
DNA	deoxyribonucleic acid
DOX	doxorubicin
DPH NGs	well-designed prodrug nanogels
ER	endoplasmic reticulum
GNPs	gold nanoparticles
GNR	gold nanorods
GO	graphene oxide
HCPT	10-hydroxycamptothecin
IC_{50}	half maximal inhibitory concentration
IL	ionic liquid
LED	light-emitting diode
MPPa	methyl pyropheophorbide
MSC	mesenchymal stem cells
NFs	nanofibers
NIR	near-infrared region
PAI	photoacoustic imaging
Pb*a*	pheophorbide *a*
PDT	photodynamic therapy
PEG	polyethylene glycol
PMTEMA	poly[2-(methylthio)ethyl methacrylate]
POEGMA	poly(oligoethylene glycol) methacrylate
PS	photosensitizer
PTT	photothermal therapy
Pu18	purpurin 18
Pu18ME	purpurin 18 methyl ester
RBCM	red blood cell memembrane
RGD-Dex	peptide RGD with dextran
ROS	reactive oxygen species
SDT	sonodynamic therapy
ST	sonosensitizer
WSONs	water-soluble organic nanoparticles
ZnMPPa	zinc metal-introduced methyl pyropheophorbide

References

1. World Health Organization. Available online: https://www.who.int/health-topics/cancer#tab=tab_1 (accessed on 28 December 2020).
2. Hopper, C. Photodynamic therapy: A clinical reality in the treatment of cancer. *Lancet Oncol.* **2000**, *1*, 212–219. [CrossRef]
3. Romano, G.; Costantini, M.; Sansone, C.; Lauritano, C.; Ruocco, N.; Ianora, A. Marine microorganisms as a promising and sustainable source of bioactive molecules. *Mar. Environ. Res.* **2017**, *128*, 58–69. [CrossRef] [PubMed]
4. Oo, Y.Y.N.; Su, M.C.; Kyaw, K.T. Extraction and determination of chlorophyll content from microalgae. *Int. J. Adv. Res. Public* **2017**, *1*, 298–301.
5. Shahidi, F.; Janak Kamil, Y.V.A. Enzymes from fish and aquatic invertebrates and their application in the food industry. *Trends Food Sci. Technol.* **2001**, *12*, 435–464. [CrossRef]
6. Pereira, D.M.; Valentão, P.; Andrade, P.B. Marine natural pigments: Chemistry, distribution and analysis. *Dyes Pigm.* **2014**, *111*, 124–134. [CrossRef]
7. Pangestuti, R.; Kim, S.K. Biological activities and health benefit effects of natural pigments derived from marine algae. *J. Funct. Foods* **2011**, *3*, 255–266. [CrossRef]
8. Holdt, S.L.; Kraan, S. Bioactive compounds in seaweed: Functional food applications and legislation. *J. Appl. Phycol.* **2011**, *23*, 543–597. [CrossRef]
9. Khan, S.B.; Kong, C.S.; Kim, J.A.; Kim, S.K. Protective effect of *Amphiroa dilatata* on ROS induced oxidative damage and MMP expressions in HT1080 cells. *Biotechnol. Bioproc. Eng.* **2010**, *15*, 191–198. [CrossRef]
10. Hosikian, A.; Lim, S.; Halim, R.; Danquah, M.K. Chlorophyll extraction from microalgae: A review on the process engineering aspects. *Int. J. Chem. Eng.* **2010**, *2010*, 391632. [CrossRef]
11. Manivasagan, P.; Bharathiraja, S.; Moorthy, M.S.; Mondal, S.; Seo, H.; Lee, K.D.; Oh, J. Marine natural pigments as potential sources for therapeutic applications. *Crit. Rev. Biotechnol.* **2018**, *38*, 745–761. [CrossRef]
12. Larkum, A.W.D.; Kühl, M. Chlorophyll d: The puzzle resolved. *Trends Plant. Sci.* **2005**, *10*, 355–357. [CrossRef] [PubMed]
13. Rüdiger, W. Biosynthesis of chlorophyll b and the chlorophyll cycle. *Photosynth. Res.* **2002**, *74*, 187–193. [CrossRef]
14. Tanaka, A.; Tanaka, R. Chlorophyll metabolism. *Curr. Opin. Plant. Biol.* **2006**, *9*, 248–255. [CrossRef] [PubMed]
15. Kuai, B.; Chen, J.; Hörtensteiner, S. The biochemistry and molecular biology of chlorophyll breakdown. *J. Exp. Bot.* **2018**, *69*, 751–767. [CrossRef] [PubMed]
16. Zvezdanović, J.; Marković, D. Bleaching of chlorophylls by UV irradiation in vitro: The effects on chlorophyll organization in acetone and n-hexane. *Serb. Chem. Soc.* **2008**, *73*, 271–282. [CrossRef]
17. Koca, N.; Karadeniz, F.; Burdurlu, S.H. Effect of pH on chlorophyll degradation and colour loss in blanched green peas. *Food Chem.* **2007**, *100*, 609–615. [CrossRef]
18. Hoober, J.K.; Sery, T.W.; Yamamoto, N. Photodynamic sensitizers from chlorophyll: Purpurin-18 and chlorin p6. *Photochem. Photobiol.* **1988**, *48*, 579–582. [CrossRef]
19. Drogat, N.; Barrière, M.; Granet, R.; Sol, V.; Krausz, P. High yield preparation of purpurin-18 from *Spirulina maxima*. *Dyes Pigm.* **2011**, *88*, 125–127. [CrossRef]
20. Watanabe, N.; Yamamoto, K.; Ihshikawa, H.; Yagi, A.; Sakata, K.; Brinen, L.S.; Clardy, J. New chlorophyll-a-related compounds isolated as antioxidants from marine bivalves. *J. Nat. Prod.* **1993**, *56*, 305–317. [CrossRef]
21. Louda, J.W.; Neto, R.R.; Magalhaes, A.R.M.; Schneider, V.F. Pigment alterations in the brown mussel *Perna perna*. *Comp. Biochem. Physiol. B Biochem. Mol. Biol.* **2008**, *150*, 385–394. [CrossRef]
22. Chen, K.; Ríos, J.J.; Pérez-Gálvez, A.; Roca, M. Comprehensive chlorophyll composition in the main edible seaweeds. *Food Chem.* **2017**, *228*, 625–633. [CrossRef] [PubMed]
23. Ocampo, R.; Repeta, D.J. Structural determination of purpurin-18 (as methyl ester) from sedimentary organic matter. *Org. Geochem.* **1999**, *30*, 189–193. [CrossRef]
24. Zenkevich, E.; Sagun, E.; Knyukshto, V.; Shulga, A.; Mironov, A.; Efremova, O.; Bonnett, R.; Songca, S.P.; Kassem, M. Photophysical and photochemical properties of potential porphyrin and chlorin photosensitizers for PDT. *J. Photochem. Photobiol. B Biol.* **1996**, *33*, 171–180. [CrossRef]
25. Zhou, J.; Gao, Z.J.; Cai, J.Q.; Li, L.L.; Wang, H. Synthesis and self-assembly behavior of chlorophyll derivatives for ratiometric photoacoustic signal optimization. *Langmuir* **2020**, *36*, 1559–1568. [CrossRef] [PubMed]
26. Zhang, D.; Qi, G.B.; Zhao, Y.X.; Qiao, S.L.; Yang, C.; Wang, H. In situ formation of nanofibers from purpurin18-peptide conjugates and the assembly induced retention effect in tumor sites. *Adv. Mater.* **2015**, *27*, 6125–6130. [CrossRef] [PubMed]
27. Reczek, C.R.; Chandel, N.S. The two faces of reactive oxygen species in cancer. *Ann. Rev. Cancer Biol.* **2017**, *1*, 79–98. [CrossRef]
28. Plaetzer, K.; Krammer, B.; Berlanda, J.; Berr, F.; Kiesslich, K. Photophysics and photochemistry of photodynamic therapy: Fundamental aspects. *Lasers Med. Sci.* **2009**, *24*, 259–268. [CrossRef]
29. Richards-Kortum, R.; Sevick-Muraca, E. Quantitative optical spectroscopy for tissue diagnosis. *Annu. Rev. Phys. Chem.* **1996**, *47*, 555–606. [CrossRef]
30. Hemmer, E.; Benayas, A.; Légaréa, F.; Vetrone, F. Exploiting the biological windows: Current perspectives on fluorescent bioprobes emitting above 1000 nm. *Nanoscale Horiz.* **2016**, *1*, 168–184. [CrossRef] [PubMed]
31. Di Stefano, A.; Ettorre, A.; Sbrana, S.; Giovani, C.; Neri, P. Purpurin-18 in combination with light leads to apoptosis or necrosis in HL60 leukemia cells. *Photochem. Photobiol.* **2001**, *73*, 290–296. [CrossRef]

32. Magi, B.; Ettorre, A.; Liberatori, S.; Bini, L.; Andreassi, M.; Frosali, S.; Neri, P.; Pallini, V.; Di Stefano, A. Selectivity of protein carbonylation in the apoptotic response to oxidative stress associated with photodynamic therapy: A cell biochemical and proteomic investigation. *Cell Death Differ.* **2004**, *11*, 842–852. [CrossRef]
33. Huang, P.; Zhang, B.; Yuan, Q.; Zhang, X.; Leung, W.; Xu, C. Photodynamic treatment with purpurin 18 effectively inhibits triple negative breast cancer by inducing cell apoptosis. *Lasers Med. Sci.* **2021**, *36*, 339–347. [CrossRef] [PubMed]
34. Zheng, G.; Potter, W.R.; SumLin, A.; Dougherty, T.J.; Pandey, R.K. Photosensitizers related to purpurin-18-*N*-alkylimides: A comparative in vivo tumoricidal ability of ester versus amide functionalities. *Bioorg. Med. Chem. Lett.* **2000**, *10*, 123–127. [CrossRef]
35. Wang, J.J.; Yin, Y.F.; Yang, Z. Synthesis of purpurin-18 imide derivatives from chlorophyll-*a* and-*b* by modifications and functionalizations along their peripheries. *J. Iran. Chem. Soc.* **2013**, *10*, 583–591. [CrossRef]
36. Wipo IP Portal. Available online: https://patentscope.wipo.int/search/en/detail.jsf?docId=WO1995032206 (accessed on 24 February 2021).
37. Lens.org. Available online: https://www.lens.org/lens/patent/US_5591847_A/fulltext (accessed on 24 February 2021).
38. Pavlíčková, V.; Rimpelová, S.; Jurášek, M.; Záruba, K.; Fähnrich, J.; Křížová, I.; Bejček, J.; Rottnerová, Z.; Spiwok, V.; Drašar, P.; et al. PEGylated purpurin 18 with improved solubility: Potent compounds for photodynamic therapy of cancer. *Molecules* **2019**, *24*, 4477. [CrossRef] [PubMed]
39. Zhang, Y.; Zhang, H.; Wang, Z.; Jin, Y. pH-Sensitive graphene oxide conjugate purpurin-18 methyl ester photosensitizer nanocomplex in photodynamic therapy. *New J. Chem.* **2018**, *42*, 13272–13284. [CrossRef]
40. Kang, E.S.; Lee, T.H.; Liu, Y.; Han, K.H.; Lee, W.K.; Yoon, I. Graphene oxide nanoparticles having long wavelength absorbing chlorins for highly-enhanced photodynamic therapy with reduced dark toxicity. *Int. J. Mol. Sci.* **2019**, *20*, 4344. [CrossRef]
41. Liu, Y.; Lee, T.H.; Lee, S.H.; Li, J.; Lee, W.K.; Yoon, I. Mitochondria-targeted water-soluble organic nanoparticles of chlorin derivatives for biocompatible photodynamic therapy. *ChemNanoMat* **2020**, *6*, 610–617. [CrossRef]
42. Liu, Y.; Lee, S.H.; Lee, W.K.; Yoon, I. Ionic liquid-dependent gold nanoparticles of purpurin-18 for cellular imaging and photodynamic therapy in vitro. *Bull. Korean Chem. Soc.* **2020**, *41*, 230–233. [CrossRef]
43. Adawiyah, N.; Moniruzzaman, M.; Hawatulaila, S.; Goto, M. Ionic liquids as a potential tool for drug delivery systems. *Med. Chem. Commun.* **2016**, *7*, 1881–1897. [CrossRef]
44. Darmostuk, M.; Jurášek, M.; Lengyel, K.; Zelenka, J.; RumLová, M.; Drašar, P.; RumL, T. Conjugation of chlorins with spermine enhances phototoxicity to cancer cells in vitro. *J. Photochem. Photobiol. B Biol.* **2017**, *168*, 175–184. [CrossRef] [PubMed]
45. Cao, B.; Yang, M.; Zhu, Y.; Qu, X.; Mao, C. Stem cells loaded with nanoparticles as a drug carrier for in vivo breast cancer therapy. *Adv. Mater.* **2014**, *26*, 4627–4631. [CrossRef]
46. Bechet, D.; Auger, F.; Couleaud, P.; Marty, E.; Ravasi, L.; Durieux, N.; Bonnet, C.; Plénat, F.; Frochot, C.; Mordon, S.; et al. Multifunctional ultrasmall nanoplatforms for vascular-targeted interstitial photodynamic therapy of brain tumors guided by real-time MRI. *Nanomedicine* **2015**, *11*, 657–670. [CrossRef] [PubMed]
47. Rai, P.; Mallidi, S.; Zheng, X.; Rahmanzadeh, R.; Mir, Y.; Elrington, S.; Khurshid, A.; Hasan, T. Development and applications of photo-triggered theranostic agents. *Adv. Drug Deliv. Rev.* **2010**, *62*, 1094–1124. [CrossRef]
48. Lustig, R.A.; Vogl, T.J.; Fromm, D.; Cuenca, R.; Hsi, A.; D'Cruz, A.K.; Krajina, Z.; Turić, M.; Singhal, A.; Chen, J.C. A multicenter phase I safety study of intratumoral photoactivation of talaporfin sodium in patients with refractory solid tumors. *Cancer* **2003**, *98*, 1767–1771. [CrossRef] [PubMed]
49. Chen, J.; Keltner, L.; Christophersen, J.; Zheng, F.; Krouse, M.; Singhal, A.; Wang, S. New technology for deep light distribution in tissue for phototherapy. *Cancer J.* **2002**, *8*, 154–163. [CrossRef]
50. Kidd, S.; Spaeth, E.; Dembinski, J.L.; Dietrich, M.; Watson, K.; Klopp, A.; Battula, V.L.; Weil, M.; Andreeff, M.; Marini, F.C. Direct evidence of mesenchymal stem cell tropism for tumor and wounding microenvironments using in vivo bioluminescent imaging. *Stem Cells* **2009**, *27*, 2614–2623. [CrossRef] [PubMed]
51. Ren, Y.J.; Zhang, H.; Huang, H.; Wang, X.M.; Zhou, Z.Y.; Cui, F.Z.; An, Y.H. In vitro behavior of neural stem cells in response to different chemical functional groups. *Biomaterials* **2009**, *30*, 1036–1044. [CrossRef]
52. Ma, X.; Zhang, T.; Qiu, W.; Liang, M.; Gao, Y.; Xue, P.; Kang, Y.; Xu, Z. Bioresponsive prodrug nanogel-based polycondensate strategy deepens tumor penetration and potentiates oxidative stress. *Chem. Eng. J.* **2020**, 127657. [CrossRef]
53. Jia, D.; Ma, X.; Lu, Y.; Li, X.; Hou, S.; Gao, Y.; Xue, P.; Kang, Y.; Xu, Z. ROS-responsive cyclodextrin nanoplatform for combined photodynamic therapy and chemotherapy of cancer. *Chin. Chem. Lett.* **2020**. [CrossRef]
54. Yumita, N.; Nishigaki, R.; Umemura, K.; Umemura, S. Hematoporphyrin as a sensitizer of cell-damaging effect of ultrasound. *Jpn. J. Cancer Res.* **1989**, *80*, 219–222. [CrossRef] [PubMed]
55. Yumita, N.; Nishigaki, R.; Umemura, S. Sonodynamically induced antitumor effect of photofrin II on colon 26 carcinoma. *J. Cancer Res. Clin. Oncol.* **2000**, *126*, 601–606. [CrossRef] [PubMed]
56. Endo, S.; Kudo, N.; Yamaguchi, S.; Sumiyoshi, K.; Motegi, H.; Kobayashi, H.; Terasaka, S.; Houkin, K. Porphyrin derivatives-mediated sonodynamic therapy for malignant gliomas in vitro. *Ultrasound Med. Biol.* **2015**, *41*, 2458–2465. [CrossRef] [PubMed]
57. Jin, Z.H.; Miyoshi, N.; Ishiguro, K.; Umemura, S.; Kawabata, K.; Yumita, N.; Sakata, I.; Takaoka, K.; Udagawa, T.; Nakajima, S.; et al. Combination effect of photodynamic and sonodynamic therapy on experimental skin squamous cell carcinoma in C_3H/HeN mice. *J. Dermatol.* **2000**, *27*, 294–306. [CrossRef]
58. Umemura, K.; Yumita, N.; Nishigaki, R.; Umemura, S. Sonodynamically induced antitumor effect of pheophorbide a. *Cancer Lett.* **1996**, *102*, 151–157. [CrossRef]

59. Singh, P.; Pandit, S.; Mokkapati, V.R.S.S.; Garg, A.; Ravikumar, V.; Mijakovic, I. Gold nanoparticles in diagnostics and therapeutics for human cancer. *Int. J. Mol. Sci.* **2018**, *19*, 1979. [CrossRef]
60. Rosenthal, I.; Sostaric, J.Z.; Riesz, P. Sonodynamic therapy-a review of synergistic effects of drugs and ultrasound. *Ultrason Snochem.* **2004**, *11*, 349–363. [CrossRef]
61. Canavese, G.; Ancona, A.; Racca, L.; Canta, M.; Dumontel, B.; Barbaresco, F.; Limongi, T.; Cauda, V. Nanoparticle-assisted ultrasound: A special focus on sonodynamic therapy against cancer. *Chem. Eng. J.* **2018**, *340*, 155–172. [CrossRef]
62. Mišík, V.; Riesz, P. Free radical intermediates in sonodynamic therapy. *Ann. N. Y. Acad. Sci.* **2000**, *899*, 335–348. [CrossRef]
63. McHale, A.P.; Callan, J.F.; Nomikou, N.; Fowley, C.; Callan, B. Sonodynamic therapy: Concept, mechanism and application to cancer treatment. In *Therapeutic Ultrasound. Advances in Experimental Medicine and Biology*; Escoffre, J.M., Bouakaz, A., Eds.; Springer International Publishing: Cham, Switzerland, 2016; Volume 880, pp. 429–450. ISBN 978-3-319-22536-4. [CrossRef]
64. Zhou, Y.; Wang, D.; Zhang, Y.; Chitgupi, U.; Geng, J.; Wang, Y.; Zhang, Y.; Cook, T.R.; Xia, J.; Lovell, J.F. A phosphorus phthalocyanine formulation with intense absorbance at 1000 nm for deep optical imaging. *Theranostics* **2016**, *6*, 688–697. [CrossRef]
65. Shi, X.; Zhang, Y.; Tian, Y.; Xu, S.; Ren, E.; Bai, S.; Chen, X.; Chu, C.; Xu, Z.; Liu, G. Multi-responsive bottlebrush-like unimolecules self-assembled nano-riceball for synergistic sono-chemotherapy. *Small Methods* **2020**, 2000416. [CrossRef]
66. Enyedi, K.N.; Tóth, S.; Szakács, G.; Mező, G. NGR-peptide−drug conjugates with dual targeting properties. *PLoS ONE* **2017**, *12*, e0178632. [CrossRef]
67. Sun, Q.; Wu, J.; Jin, L.; Hong, L.; Wang, F.; Mao, Z.; Wu, M. Cancer cell membrane-coated gold nanorods for photothermal therapy and radiotherapy on oral squamous cancer. *J. Mater. Chem. B* **2020**, *8*, 7253–7263. [CrossRef] [PubMed]
68. Cheng, D.B.; Zhang, X.H.; Chen, Y.; Chen, H.; Qiao, Z.Y.; Wang, H. Ultrasound-activated cascade effect for synergistic orthotopic pancreatic cancer therapy. *iScience* **2020**, *23*, 101144. [CrossRef]
69. Attia, A.B.E.; Balasundaram, G.; Moothanchery, M.; Dinish, U.S.; Bi, R.; Ntziachristos, V.; Olivo, M. A review of clinical photoacoustic imaging: Current and future trends. *Photoacoustics* **2019**, *16*, 100144. [CrossRef]
70. Clinical Trials. Available online: https://clinicaltrials.gov/ct2/show/NCT03897270 (accessed on 28 December 2020).
71. Clinical Trials. Available online: https://clinicaltrials.gov/ct2/show/NCT04428528?term=photoacoustic&cond=Cancer&draw=2&rank=8 (accessed on 28 December 2020).
72. Clinical Trials. Available online: https://clinicaltrials.gov/ct2/show/NCT04110249?term=NCT04110249&draw=2&rank=1 (accessed on 28 December 2020).
73. Available online: https://clinicaltrials.gov/ct2/show/NCT04437030?term=photoacoustic&cond=Cancer&draw=2&rank=9 (accessed on 28 December 2020).
74. Clinical Trials. Available online: https://clinicaltrials.gov/ct2/show/NCT04248166?term=photoacoustic&cond=Cancer&draw=2&rank=1 (accessed on 28 December 2020).
75. Clinical Trials. Available online: https://clinicaltrials.gov/ct2/show/NCT04339374?term=photoacoustic&cond=Cancer&draw=2&rank=6 (accessed on 28 December 2020).
76. Kim, C.; Song, K.H.; Gao, F.; Wang, L.V. Sentinel lymph nodes and lymphatic vessels: Noninvasive dual-modality in vivo mapping by using indocyanine green in rats—volumetric spectroscopic photoacoustic imaging and planar fluorescence imaging. *Radiology* **2010**, *255*, 442–450. [CrossRef]
77. Song, K.H.; Stein, E.W.; Margenthaler, J.A.; Wang, L.V. Noninvasive photoacoustic identification of sentinel lymph nodes containing methylene blue in vivo in a rat model. *J. Biomed. Opt.* **2008**, *13*, 054033. [CrossRef]
78. Zhang, D.; Wang, Z.; Wang, L.; Wang, Z.; Wang, H.; Li, G.; Qiao, Z.Y.; Xu, W.; Wang, H. High-performance identification of human bladder cancer using a signal self-amplifiable photoacoustic nanoprobe. *ACS Appl. Mater. Interfaces* **2018**, *10*, 28331–28339. [CrossRef]
79. Zhao, L.H.; Lin, Q.L.; Wei, J.; Huai, Y.L.; Wang, K.J.; Yan, H.Y. CD44v6 expression in patients with stage II or stage III sporadic colorectal cancer is superior to CD44 expression for predicting progression. *Int. J. Clin. Exp. Pathol.* **2015**, *8*, 692–701. [PubMed]
80. Ma, L.; Dong, L.; Chang, P. CD44v6 engages in colorectal cancer progression. *Cell Death Dis.* **2019**, *10*, 30. [CrossRef] [PubMed]
81. Hu, S.; Cao, M.; He, Y.; Zhang, G.; Liu, Y.; Du, Y.; Yang, C.; Gao, F. CD44v6 targeted by miR-193b-5p in the coding region modulates the migration and invasion of breast cancer cells. *J. Cancer* **2020**, *11*, 260–271. [CrossRef]
82. Lipponen, P.; Aaltoma, S.; Kosma, V.M.; Ala-Opas, M.; Eskelinen, M. Expression of CD44 standard and variant-v6 proteins in transitional cell bladder tumours and their relation to prognosis during a long-term follow-up. *J. Pathol.* **1998**, *186*, 157–164. [CrossRef]
83. Zhang, X.H.; Cheng, D.B.; Ji, L.; An, H.W.; Wang, D.; Yang, Z.X.; Chen, H.; Qiao, Z.Y.; Wang, H. Photothermal-promoted morphology transformation in vivo monitored by photoacoustic imaging. *Nano Lett.* **2020**, *20*, 1286–1295. [CrossRef]
84. Cheng, D.B.; Qi, G.B.; Wang, J.Q.; Cong, Y.; Liu, F.H.; Yu, H.; Qiao, Z.Y.; Wang, H. In situ monitoring intracellular structural change of nanovehicles through photoacoustic signals based on phenylboronate-linked RGD-dextran/purpurin 18 conjugates. *Biomacromolecules* **2017**, *18*, 1249–1258. [CrossRef]
85. Xiao, B.; Hong, L.; Cai, X.; Mei, S.; Zhang, P.; Shao, L. The true colors of autophagy in doxorubicin-induced cardiotoxicity. *Oncol. Lett.* **2019**, *18*, 2165–2172. [CrossRef]
86. Chen, C.; Lu, L.; Yan, S.; Yi, H.; Yao, H.; Wu, D.; He, G.; Tao, X.; Deng, X. Autophagy and doxorubicin resistance in cancer. *Anti-Cancer Drugs* **2018**, *29*, 1–9. [CrossRef] [PubMed]
87. Mulcahy Levy, J.M.; Thorburn, A. Autophagy in cancer: Moving from understanding mechanism to improving therapy responses in patients. *Cell Death Diff.* **2020**, *27*, 843–857. [CrossRef] [PubMed]

88. Lin, Y.X.; Wang, Y.; Qiao, S.L.; An, H.W.; Wang, J.; Ma, Y.; Wang, L.; Wang, H. "In vivo self-assembled" nanoprobes for optimizing autophagy-mediated chemotherapy. *Biomaterials* **2017**, *141*, 199–209. [CrossRef] [PubMed]
89. Liu, W.J.; Zhang, D.; Li, L.L.; Qiao, Z.Y.; Zhang, J.C.; Zhao, Y.X.; Qi, G.B.; Wan, D.; Pan, J.; Wang, H. In situ construction and characterization of chlorin-based supramolecular aggregates in tumor cells. *ACS Appl. Mater. Interfaces* **2016**, *8*, 22875–22883. [CrossRef] [PubMed]
90. Muz, B.; de la Puente, P.; Azab, F.; Azab, A.K. The role of hypoxia in cancer progression, angiogenesis, metastasis, and resistance to therapy. *Hypoxia* **2015**, *3*, 83–92. [CrossRef] [PubMed]
91. Kim, S.B.; Lee, T.H.; Yoon, I.; Shim, Y.K.; Lee, W.K. Gold nanorod–photosensitizer complex obtained by layer-by-layer method for photodynamic/photothermal therapy in vitro. *Chem. Asian J.* **2015**, *10*, 563–567. [CrossRef] [PubMed]
92. Turner, J.H. An introduction to the clinical practice of theranostics in oncology. *Br. J. Radiol.* **2018**, *91*, 20180440. [CrossRef] [PubMed]
93. Sun, Y.; Zhang, Y.; Gao, Y.; Wang, P.; He, G.; Blum, N.T.; Lin, J.; Liu, Q.; Wang, X.; Huang, P. Six birds with one stone: Versatile nanoporphyrin for single-laser-triggered synergistic phototheranostics and robust immune activation. *Adv. Mater.* **2020**, *32*, 2004481. [CrossRef] [PubMed]

Article

Characterization of Alginate from *Sargassum duplicatum* and the Antioxidant Effect of Alginate–Okra Fruit Extracts Combination for Wound Healing on Diabetic Mice

Zulfa Nailul Ilmi [1], Pugar Arga Cristina Wulandari [1], Saikhu Akhmad Husen [2], Dwi Winarni [2], Mochammad Amin Alamsjah [3], Khalijah Awang [4], Marco Vastano [5], Alessandro Pellis [5,6], Duncan Macquarrie [5] and Pratiwi Pudjiastuti [1,*]

1 Department of Chemistry, Faculty of Science and Technology, Airlangga University, Surabaya 60115, Indonesia; zulfanailulilmi@gmail.com (Z.N.I.); cristinapugar@gmail.com (P.A.C.W.)
2 Department of Biology, Faculty of Science and Technology, Airlangga University, Surabaya 60115, Indonesia; saikhu-a-h@fst.unair.ac.id (S.A.H.); dwi-w@fst.unair.ac.id (D.W.)
3 Department of Marine, Faculty of Fisheries and Marine, Airlangga University, Surabaya 60115, Indonesia; alamsjah@fpk.unair.ac.id
4 Department of Chemistry, Faculty of Science, University of Malaya, Kuala Lumpur 50603, Malaysia; khalijah@um.edu.my
5 Department of Chemistry, University of York, Heslington, York YO10 5DD, UK; marco.vastano@york.ac.uk (M.V.); alessandro.pellis@boku.ac.at (A.P.); duncan.macquarrie@york.ac.uk (D.M.)
6 Department for Agrobiotechnology, IFA-Tulln, Institute for Environmental Biotechnology, University of Natural Resources and Life Sciences, Vienna, Konrad Lorenz Strasse 20, A-3430 Tulln an der Donau, Austria
* Correspondence: pratiwi-p@fst.unair.ac.id; Tel.: +62-856-3390-952

Received: 12 July 2020; Accepted: 27 August 2020; Published: 2 September 2020

Featured Application: In this study, we evaluated antioxidant effect of alginate–okra extracts combination as topical application (ointment) for wound healing on diabetic mice. Characterization of alginate was used to determine structural characteristics and Okra fruit extract was determined total flavonoids content that have effect of the antioxidant properties. The antioxidant properties of extracts combination reduce blood-glucose levels to non-diabetic conditions (normal) significantly by topical application of diabetic open wound. These conditions can accelerate the activities of wound-healing processes on diabetic mice. The activities of wound-healing processes were performing periodically by histopathology test on skin tissue to evaluated wound healing parameters (wound area, neutrophils, macrophages, fibrocytes, fibroblasts and collagen densities).

Abstract: Diabetes mellitus is a group of metabolic disorders characterized by high blood-glucose levels over a prolonged period that causes complications when an open wound is present. Alginate is an antioxidant and a good absorbent of exudates. Okra fruit contains flavonoids that can act as antioxidants. The antioxidant properties of extracts combination reduce blood-glucose levels significantly to accelerate the activities of wound-healing processes on diabetic mice. Alginate was characterized by Size Exclusion Chromatography-Multiple Angle Laser Light Scattering (SEC-MALLS), thermal stability and Proton Nuclear Magnetic Resonance (^{1}H-NMR). The evaluation of wound healing on 36 male mice were divided into 12 groups including normal control (NC), diabetics control (DC), alginate (DA) and alginate–okra (DAO) groups in three different times by histopathology test on skin tissue. The results of SEC-MALLS analysis showed that alginate as single and homogeneous polysaccharide. The ^{1}H-NMR spectrum showed that the mannuronate/guluronate ratio of the used alginate was 0.91. Alginate, okra fruit extract and their combination were classified as moderate and strong antioxidants. The numbers of fibrocytes, fibroblasts, collagen densities had significantly increased from three to seven days. In contrast, wound width, neutrophil, macrophages had significantly decreased at 14 days. The administration of extracts combination increased the re-epithelization of the wound area and wound-healing process on diabetic mice.

Keywords: alginate; *Sargassum duplicatum*; okra; antioxidant; diabetic; wound healing

---

1. Introduction

Diabetes mellitus (DM) is a disease that causes complex metabolic disorder associated with insulin resistance, impaired insulin signaling, β-cell dysfunction, abnormal glucose levels and changes in lipid metabolism [1]. The condition of high blood-glucose level (hyperglycemia) is caused by impaired of insulin production or secretion, insulin resistance. Hyperglycemia causes an increase of reactive oxygen species (ROS) levels which is not counterbalanced by an adequate supply of antioxidants. This imbalance can cause oxidative stress which is the source of the development of diabetes mellitus complications because it triggers the development of insulin resistance [2]. Complication of diabetes mellitus causes chronic wound healing disorders such as inflammation, tissue formation granulation, re-epithelization and tissue renovation [3].

All these complications often result in open wounds with long-term healing problems that frequently cause the need of amputation. Epidemiological data report that 2.8% of the world's population suffered from diabetes in 2000, and the trend is expected to increase up to 4.4% by 2030 and also the percentage of amputation in diabetic wounds is expected to increase to about 50–70% [4]. The excessive generation of ROS in diabetic conditions is caused by acute serum glucose and the accumulation of advanced glycation end-products (AGEs). Endogenous or externally supplied antioxidants (due to inadequate supply of antioxidants) can lower ROS levels through the reduction of cellular molecular oxidation. Antioxidants obtained can neutralize the oxidative stress to improve wound healing in diabetes [5].

Wound-healing processes in diabetic mice have been usually treated performed in general through the administration of topical medications, topical or systemic antibiotics and vascular reconstruction. However, open wound healing has a weakness due to the use of chemical substances that cause long-term side effects in diabetics [6,7].

Alginate is a natural polysaccharide produced from *Sargassum* sp. (among others) containing α-L-guluronic acid (G) and β-D-mannuronic acid (M) residues [8]. Alginate—in form of hydrogels, foams, films, nanofibers, topical formulations or wafers—has been widely used as bioactive dressing for wound healing [9]. As other biocompatible materials, alginate can be loaded natural compound or drugs for improved wound-healing process like simvastatin [10], honey [11], *Aloe vera* [12], curcumin [13]. Alginate can maintain a moist condition in the wound area and acts as antibacterial and antioxidant agent, absorb wound exudate as a hemostatic agent [14–16]. Moreover, alginate with composition of guluronic acid residue (G) higher than Mannuronic acid residue (M) produce a more rigid structure, thereby increasing the availability of alginate ions to make it easier to absorb exudate and proton donor in the free radical scavenging process (antioxidant), so it is easy to wound healing. Molecular weight of polysaccharides affects the ability of radical scavenging. Reduction in molecular weight below 12 kDa can affect the conformational chain or guluronic acid content which has an impact on the availability of alginate ions in radical scavenging [17].

Abelmoschus esculentus L. is known as okra plant that contains many flavonoids that play a major role acting as antioxidant agents against ROS to inhibit free radicals [18]. Hence, the okra fruit may be used as an alternative medicine for the reduction of blood glucose and cholesterol levels as antidiabetic and antihyperlipidemic properties [19,20]. In addition, okra fruit extract can accelerate wound healing through the induction of the production of cytokine and growth factor productions [21].

In this study, we report the characterization of alginate from *Sargassum duplicatum* (J. Agardh) to evaluate its characteristic on the wound-healing process. The novelty of this study lies in investigating antioxidant effect of alginate from *S. duplicatum* and okra fruit extracts combination for wound-healing process in diabetic mice. The activities of wound-healing process were performed by using an alginate–okra extracts ointment on wound performing periodically by histopathology tests on diabetic mice skin tissues.

2. Materials and Methods

2.1. Plants Collection

Brown seaweeds of *S. duplicatum* were taken from Madura Island, Indonesia in May 2018. The species was identified at Oceanographic Research Center, LIPI, Jakarta, Indonesia (Approval Reference Number: 1/3/18-id/2018). Okra (*A. esculentus*) fruits were obtained from Rungkut district, Surabaya, Indonesia. Identification of species was conducted at the Laboratory Biosystem of Biology Department, Airlangga University, Surabaya, Indonesia.

2.2. Extraction of Sodium Alginate

The method of extraction of alginate from *S. duplicatum* referred to research from Great Hall of Products and Biotechnology of Marine and Fisheries, Jakarta with modification from LIPI. Sample of *S. duplicatum* was cleaned, dried and grinded. Two hundrd grams of dried sample was soaked using 0.1% (1:75 *w/v*) KOH for two hours, then added of 1% (1:30 *w/v*) HCl was added and let it react for an hour. The residue was then washed with water to a neutral pH and added 2% (1:30 *w/v*) Na_2CO_3 at 60–70 °C for two hours with constant stirring. The extract was added 10% of HCl until a pH in the 2.8–3.2 range was reached. Two percent Na_2CO_3 at pH 7 was added until the formation of a gel filtrate was obtained. Gel filtrate was bleached with 4% (1:2 *v/v*) NaOCl for 15 h. Filtrate was refined by using isopropyl alcohol for 30 min to form fibers [22]. The extraction of sodium alginate powder from *S. duplicatum* was pale yellow color and 12.140 ± 0.684% yield based on the dry weight of brown seaweed.

2.3. Characterization of Sodium Alginate

The characterization of sodium alginate was conducted using the SEC-MALLS for determination of molecular weight of alginate. The C/M ratio of sodium alginate from *S. duplicatum* was identified by ^{1}H-NMR spectroscopy. The thermogravimetric analysis (TGA) was performed to evaluate the strength and thermal stability of the polymer matrix by recording phase transitions and the degradation pattern of alginate based on percent of weight loss and its decomposition under nitrogen gas. All measurements were conducted at the Department of Chemistry, University of York, Heslington, UK.

2.3.1. SEC-MALLS

The molecular weight of pectin and its distribution were determined using the gel permeation chromatography technique using a Shimadzu HPLC system (Shimadzu UK Limited, Milton Keynes, UK) comprising a CBM-20A Controller, LC-20AD Pump with degasser, SIL-20A Autosampler and SPD-20A detector; and HELEOS-II light scattering and Optilab rEx refractive index detectors supplied by Wyatt. A PL Aquagel-OH mixed column (7.5 mm × 300 mm, 8-µm particle size; Agilent Technologies, Santa Clara, CA, USA) and a PL Aquagel-OH mixed guard column (7.5 × 50 mm, 8-µm particle size; Agilent Technologies, USA) were used as the stationary phase with 50-mM sodium nitrate dissolved in deionized water as mobile phase constituting. The flow rate of the mobile phase was 0.5 mL min^{-1} [23]. The alginate solution (1.5 mg mL^{-1}), dissolved in deionized water, were filtered through a nylon membrane (Whatman, UK) before analysis. For the dn dc^{-1} value, 0.168 was used [24].

2.3.2. Nuclear Magnetic Resonance (NMR) Spectroscopy

^{1}H-NMR spectroscopy analysis was performed on a JEOL JNM-ECS400A spectrometer (JEOL, Peabody, MA, USA) at a frequency of 400 MHz for ^{1}H. D_2O was used as NMR solvent if not otherwise specified. All samples were freeze dried before analysis [25]. Acquisition parameter were: T = 90 °C and 128 scans. samples were not pre-hydrolyzed like in reference [25] since the alginate was fully soluble in D_2O.

2.3.3. Thermogravimetric Analysis (TGA)

TGA was performed on a PL Thermal Sciences STA 625 thermal analyzer (PL Thermal Sciences Limited, Surrey, UK).–10 mg of accurately weighed sample in an aluminum sample cup was placed into the furnace with a N_2 flow of 100 mL min^{-1} and heated from room temperature to 625 °C at a heating rate of 10 °C min^{-1}. From the TGA profiles the temperatures at 10% and 50% mass loss (TD_{10} and TD_{50}, respectively) were subsequently determined [26].

2.4. *Extraction of Okra Fruits*

Green okra fruits were cut, dried and milled until a fine powder was obtained. The sample was soaked by using 96% ethanol (1:3 (*w/v*)) and the procedure was thoroughly repeated [27]. The filtrate was evaporated under vacuum at 50 °C and then freeze dried. The sample was kept at 4 °C until further use.

2.5. *Determination of Total Flavonoids Content*

The total flavonoids of the okra fruits extract were determined by colorimetric aluminum chloride using UV-vis spectrophotometer [27]. Five milligrams of extract was added 5 mL 96% of ethanol until 10 mL volume. Sample solution (1 mL) was put in a 10-mL volumetric flask and added 1 mL of 10% $AlCl_3$ and 8 mL of 5% potassium acetate solution. The mixtures were incubated for 30 min at room temperature and the absorbance was determined at 431 nm. The calibration curve was made from the standard in concentrations variation of 0.01-, 0.02-, 0.03-, 0.04- and 0.05-mg/mL quercetin, respectively. The concentration variations of samples were made as same as standard of quercetin. The standard curve was plotted between concentration and absorbance, so that the equation obtained linear regression and expressed as mg quercetin equivalent (QE)/g okra fruit extract.

2.6. *Antioxidant Assay*

Antioxidant assay were performed on alginate and okra fruit extracts and their combination using DPPH method. The DPPH solution was prepared as 50.0-µg/mL concentration in methanol and extracts stock solutions were prepared as 1000.0-µg/mL concentration in methanol. The extract stock solutions were diluted in various of concentrations in methanol as 200.0-, 150.0-, 125.0-, 100.0-, 75.0-, 50.0-, 35.0-, 25.0-, 15.0-, 10.0- and 6.0-µg/mL concentrations. Dilution series of the samples as 100 µL and 100 µL of DPPH solution were added into 96-well plates. The mixtures were incubated in dark chamber for 30 min and was determined the absorbance at 517 nm by using an ELISA reader [28]. The radical DPPH scavenging activity was expressed as radical DPPH scavenging percentage using the formula:

$$\text{Radical DPPH scavenging (\%)} = \frac{A_{DPPH} - A_{sample}}{A_{DPPH}} \times 100\%$$

where, A_{DPPH} was absorbance of DPPH solution and A_{sample} was absorbance of DPPH solution in the presence of samples. IC_{50} values of the extracts were calculated from graphs plotting of radical DPPH scavenging (%) and concentration of the sample solutions [29].

2.7. *Preparation and Induction of the Diabetic Mice*

Mice (*Mus musculus*) adult male, BALB/c strains having 3–4 months of age and a body weight in the 20–35-g range were purchased from Faculty of Pharmacy, Airlangga University, Surabaya. The ethical clearance for treating animals and all the experimental protocol were approved by Faculty of Veterinary Medicine, Airlangga University, Surabaya, Indonesia with Approval Reference Number: 2. KE. 049.04.2019 (4 April 2019). Mice were acclimated at 12-h light and 12-h dark lighting system for two weeks. Food and beverage were guarded during research. Mice were then fed with lard for three weeks at a dose of 0.3 mL (oral) to obtain high fat diet. Mice were induced streptozotocin (STZ) using multiple low dose method 30 mg/kg body weight as much as 0.15 mL for 8 days at intraperitoneal (i.p.) to induce Type II diabetes mellitus [30]. The body weight of mice was measured before and after the administration of lard and STZ induction. Meanwhile, the blood-glucose levels were determined after STZ induction and after 3, 7 and 14 days of treatments.

2.8. Animal Grouping and Treatments

The thirty-six mice were divided into 12 groups: three and nine groups for non-diabetic as normal (N) and diabetic (D) groups, respectively. Each group contains three mice for 3, 7- and 14-day observation [24]. Non-diabetic mice were divided into 3 groups as normal control (NC_3, NC_7 and NC_{14}), while diabetic mice are divided into nine groups: three of diabetic control groups (DC_3, DC_7 and DC_{14}), three of alginate treatment groups (DA_3, DA_7 and DA_{14}) and three of alginate–okra treatment groups (DAO_3, DAO_7 and DAO_{14}) were given the Vaseline, Vaseline–alginate and Vaseline–alginate–okra ointments, respectively. The ointment was homogenized as single dose of 50 mg/kg body weight. The replication of the animal was based on the formula of Federer (1967) [31]. Mice were wounded to give 1-cm length wounds on the *glutea* (buttocks).

2.9. Wound Healing Observation

The histopathology preparation of skin tissue was conducted based on the method which was performed by the laboratory of Veterinary Pathology of Airlangga University. The histopathology test was observed in some parameters of wounds such as number of neutrophils, macrophages, fibrocytes, fibroblasts, collagen densities and cell regeneration areas (wound width) using a light microscope and the ImageJ software. Neutrophils were recognized as polymorphonuclear cells, lobulated nucleus which have small granules in their cytoplasm. Macrophages in connective tissue of healing area derived from monocytes which were identified as large, round cells with round nucleus and often have vacuolated cytoplasm because of their phagocytic activity. Fibroblasts are large flat and irregular shape cells, containing an oval prominent shaped nucleus and basophilic cytoplasm. In contrast, fibrocytes are inactive form of fibroblast which have darkly stained elongated nuclei, spindle shape and acidophilic cytoplasm.

2.10. Statistically Analysis

All the observations of wound healing were presented as a mean ± standard error mean (SEM). The statistical analysis of wound healing observations was performed using normality and homogeneity test, one-way ANOVA and Duncan test. Nonparametric of Kruskal–Wallis and Mann–Whitney test were used if the results of normality and homogeneity were not qualified ($\alpha = 0.05$). The difference between the various mean were calculated by using IBM company (Corporate headquarters 1 New Orchard Road Armonk, New York 10504-1722, United States US: 914-499-1900) with the SPSS 20.0 software.

3. Results

3.1. Characterization of Sodium Alginate

3.1.1. Analysis of SEC-MALLS

Sodium alginate from *S. duplicatum* was characterized by SEC-MALLS to determine molecular weight and polydispersity index. The average molecular weight of alginate *S. duplicatum* was determined via SEC-MALLS analysis (Figure 1a). Sodium alginate extracted from *S. duplicatum* contains polysaccharides with weight-average molecular weight (M_w) and number-average molecular weight (M_n) of 3.535×10^4 g/mol and 2.015×10^4 g/mol, respectively. Molecular weight of commercial alginate products from brown seaweed about 0.32–4.0 × 10^5 g/mol [32]. Alginate's dispersity (M_w/M_n) was 1.76, showing a good homogeneity of the polysaccharide alginate from *S. duplicatum*.

3.1.2. Thermogravimetric Analysis (TGA)

TGA was required to check the thermal stability and strength of the polymer matrix. TGA was used to determine the degradation temperature and the overall thermal stability of alginate. Thermal stability and strength of the polymer matrix of the alginate from *S. duplicatum* was determined via thermogravimetric analysis (Figure 1b). The 5% of the mass was lost (TD_5) at 75 °C, it indicated that water loss. The 10% of the lost mass (TD_{10}) occurred at a temperature of 140 °C. The temperature when 20% of the mass was lost (TD_{20}) occurs at 223 °C. That is indicated as the destruction of the glycosidic bond [33]. The 50% of the lost mass (TD_{50}) occurred at 380 °C. That is indicated as the material of carbon intermediates charcoal [34].

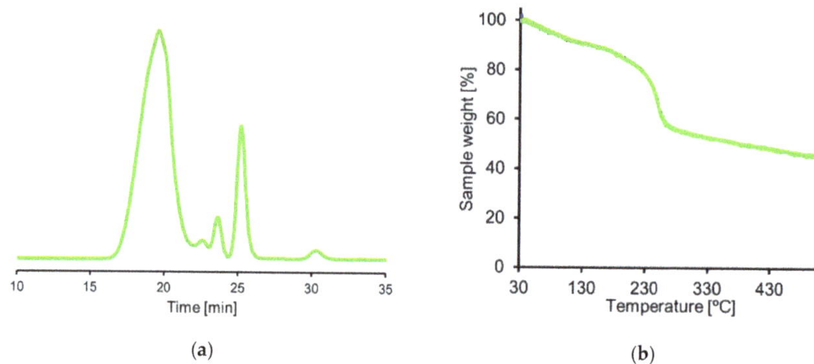

Figure 1. (**a**) Size Exclusion Chromatography-Multiple Angle Laser Light Scattering (SEC-MALLS) profile and (**b**) TGA of the sodium alginate from *Sargassum duplicatum*.

3.1.3. Proton Nuclear Magnetic Resonance (^{1}H-NMR) Analysis

The analysis was done according to Llanes et al. [25]. ^{1}H NMR spectra analysis was used to determine the Mannuronate/Guluronate (M/G) ratio of the used sodium alginate. Quantitatively, the guluronic and mannuronic mole fractions were associated with the area signal or integration of the respective peaks. The integration obtained is shown in I_A (guluronate peak area of G_1) 5.5–5.7 ppm, I_B (mannuronate peak area of M_1) 5.1–5.4 ppm, I_C (guluronic peak area of G_5) 4.9–5.1 ppm. The mole fraction of each composition of F_M and F_G is used to determine the M/G ratio of alginate with the equation referring to the research of Grasdalen et al. [35]. The signals of proton for alginate at ^{1}H-NMR spectrum has in range of 1–6 ppm. The anomer protons of each monosaccharide could be recognized from the α-glycosides and β-glycosides of the residues. Based on the ^{1}H-NMR spectra (Figure 2b), the anomer protons of α-guluronate and β-mannuronate were shown on the G_1 and M_1 proton at δ 5.6 and 5.1 ppm, respectively. Both signals are the most deshielded proton because both protons are attached to anomeric carbon. The signals of 5–6 ppm belong to anomeric proton of the α-glycosidic or α-guluronate and the signals between in the range of 4.3–5 ppm belong to proton of β-glycosidic or β-mannuronate [36]. The proton of α-glycosidic is more deshielded than β-glycosidic because it is located at equatorial position as compared to axial position. Llanes et al. [25] showed proton signals of sodium alginate from *Sargassum* sp. that G_1, M_1, G_5 proton at δ 5.07, 4.68 and 4.45 ppm. The M/G ratio of the mannuronic and guluronic acid residues can be determined from the integration of the respective peaks of the ^{1}H-NMR spectra [37]. The M/G ratio and η value of alginate *S. duplicatum* were 0.91 and 0.68, respectively. The chemical structure of mannuronic acid and guluronic acid could be shown in Figure 2c,d.

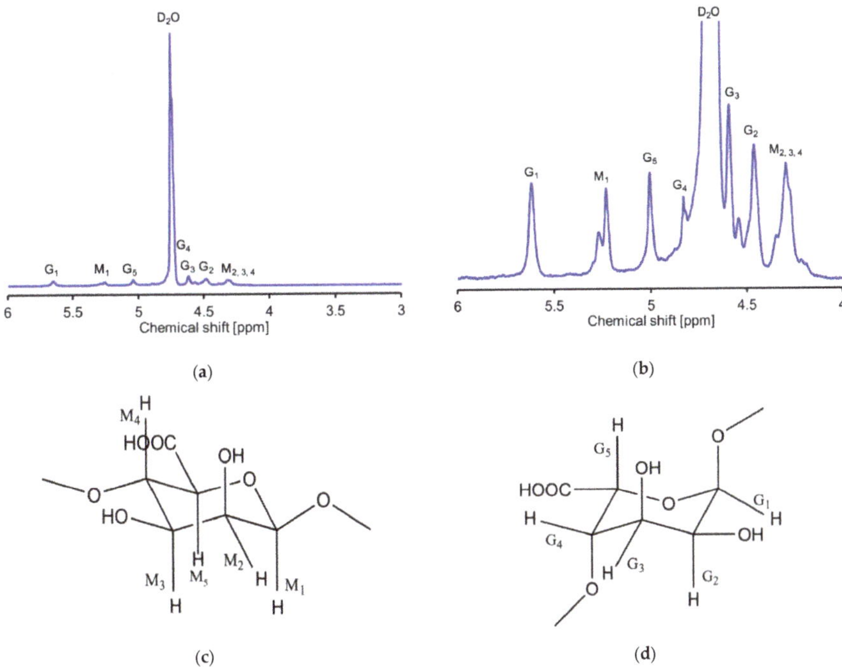

Figure 2. (**a**) ^{1}H-NMR spectra of sodium alginate from *S. duplicatum*; (**b**) details of ^{1}H-NMR spectra; (**c**) structure of mannuronic acid; (**d**) structure of guluronic acid.

3.2. Total Flavonoids Content and Antioxidant Assay

The total content of flavonoids of okra fruit extract was 43.96 mg (QE)/g or 4.40% (y = 0.0105x − 0.0161; R^2 = 0.9989). The parameter used to study the antioxidant activity is the value of inhibition concentration (IC_{50}) and it is classified in to 5 groups (very strong/highly active, strong/active, moderate, weak, inactive) [38]. In this study, ascorbic acid was selected as reference antioxidant (strong antioxidant). IC_{50} was determined using DPPH method based on the percentage of DPPH radical scavenging activity. The IC_{50} value of alginate *S. duplicatum* was 125.31-µg/mL (y = 0.2657x + 1.7814; R^2 = 0.959). Okra fruit extract and alginate–okra fruit extracts combinations (AOEs) showed strong antioxidant comparable to ascorbic acid. As reported in Table 1 the IC_{50} value of the okra fruits and of alginate–okra fruits extracts combination were 65.87-µg/mL (y = 0.097x + 43.611; R^2 = 0.9567) and 79.34-µg/mL (y = 0.3384x + 13.957; R^2 = 0.9681), respectively.

Table 1. Antioxidant activity of alginate, okra fruit and alginate–okra fruit extracts combinations (AOEs).

Sample of Extracts	IC_{50} (µg/mL)	Classification
Sodium alginate *S. duplicatum*	125.31	Moderate antioxidant
Okra fruit	65.87	Strong antioxidant
Alginate–okra extracts combination	79.34	Strong antioxidant
Ascorbic acid	55.89	Strong antioxidant

3.3. Determination of Weight on Mice

The results showed that the administration of lard for 3 weeks with a dose of 0.3 mL per oral could increase the weight of the body significantly ($\alpha < 0.05$) from 27.26 ± 4.23 g to 33.25 ± 2.97 g. Obesity of mice was characterized by increased weight after the administration of lard. (Data Not Shown)

3.4. Determination of Blood-glucose Levels

Alteration in blood-glucose levels on Day 1 to 14 after STZ induction and topical treatment is shown in Table 2. STZ was able to increase blood-glucose levels > 250 mg/dL on Day 1 which showed all diabetic groups in suffering of diabetes. Alginate administration did not lead to significantly lower blood-glucose levels ($\alpha > 0.05$) from Day 1 (251.50 ± 7.78 mg/dL) until Day 7 (229.50 ± 7.78 mg/dL) but decreased significantly at Day 14 (208.00 ± 8.49 mg/dL) if compared to the first day. In contrast, the topical treatment of alginate–okra could decrease the glucose levels significantly ($\alpha < 0.05$) from Day 1 (256.50 ± 19.09 mg/dL) until Day 14 (115.00 ± 19.06 mg/dL).

Table 2. Total blood glucose of mice on Days 1, 3, 7 and 14 after treatment.

Group Treatment	Blood-glucose Level (mg/dL)			
	Day 1	**Day 3**	**Day 7**	**Day 14**
Normal (NC)	111 ± 9.9	110.5 ± 7.8	107.5 ± 5.0	120.0 ± 14.1
Diabetic (DC)	269.5 ± 16.3	252.5 ± 19.1	287.5 ± 14.9	220.0 ± 24.0
Alginate ointment (DA)	251.5 ± 7.8	239.5 ± 27.6	229.5 ± 7.8	208.0 ± 8.5
Alginate–okra ointment (DAO)	256.5 ± 19.1	169.5 ± 24.8	182.5 ± 13.4	115.0 ± 19.1 *

Letter notation on the table shows the result of the statistical test ($\alpha < 0.05$). (*) notation shows insignificant differences with NC, but significant differences with DC. * = $0.01 \leq \alpha \leq 0.05$. n = 3 animals per time point.

3.5. Histological Analysis of Wound Tissue

Histological analysis of the width of the wound was indicated by yellow line an area of re-epithelization on the wound (Figure 3). The re-epithelization area increased on Day 7 for both DA and DAO and wounds closed occurs on Day 14 for each group. The DAO groups showed that complete re-epithelization and faster than other groups and proved by histology on Day 7 and 14.

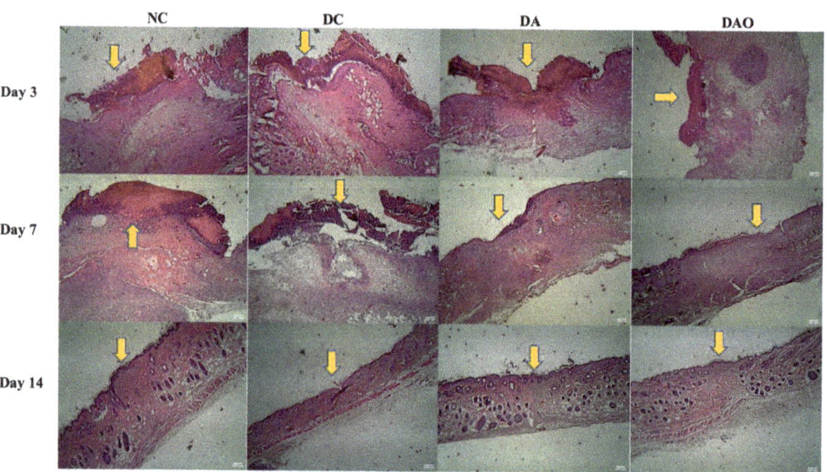

Figure 3. Histology of the wound width indicated by the yellow arrow. NC—normal control group; DC—diabetic control group; DA—alginate treatment group; DAO—treatment of alginate–okra fruit extracts. All groups have three variation times.

Wound parameter such as neutrophils (pale green), macrophages (yellow), fibrocytes (green), fibroblasts (blue) and collagen densities (orange) showed by histology magnification of wound width (Figure 4). The number of neutrophils reaches the maximum at Day 3 while it decreases from Day 3 to Day 7 and 14 for all studied groups. The number of macrophages reaches the maximum at Day 7 while it decreased at Day 14 for all groups. The DC group was evidenced by the highest concentration of macrophages when compared to other groups. Fibrocyte and fibroblast were found in the Day 7 for re-epithelization and decreased in the Day 14, as they have formed collagen fibers for all groups. The DC group was found a number of fibrocytes, fibroblasts and the least formation of collagen fibers in the group.

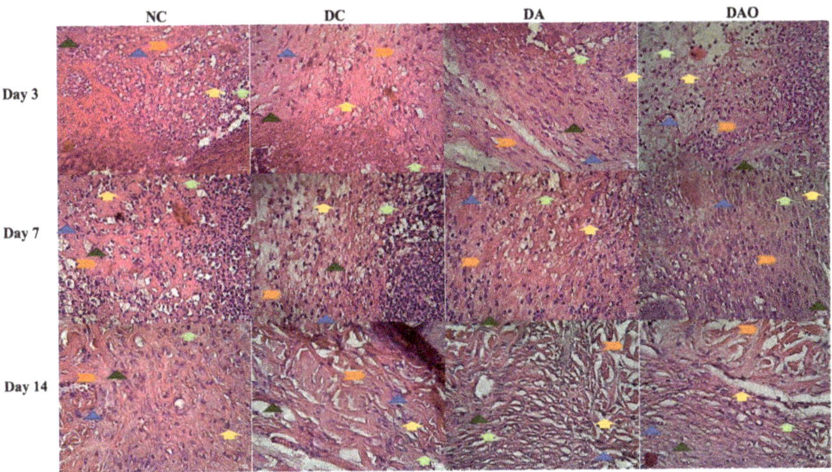

Figure 4. Histology of wound parameter including neutrophils—; Macrophages—; Fibrocytes—▲; Fibroblasts—▲; Collagen density—; NC—normal control group; DC—diabetic control group; DA—alginate treatment group; DAO—treatment of alginate–okra fruit extracts. All groups have three variation times.

3.6. Analysis of Wound Parameter

3.6.1. Wound-Width Measurement

The difference in the wound width parameter of each group for time variation is shown in Figure 5. The wound width of diabetic DC group was significantly decreased from Days 3 to 14. DC group showed significant difference to normal NC group and diabetic groups as DA, DAO treated with alginate and okra fruit extracts ointment ($\alpha < 0.05$). Treatment group of DA and DAO were significantly decrease on Days 3 to 14 and they showed significant difference on Days 3 to 7. It showed that okra fruit extracts combination plays an important role in wound closure. No significant differences between DAO group and normal group (NC) on Days 3, 7 and 14 ($\alpha > 0.05$) was observed. This indicates that the topical administration of alginate and okra fruit extracts combination can increase re-epithelization of the wound area in comparison to the control group. Diabetic group (DC) showed that still has wound width on Day 14. Otherwise, NC, DA and DAO groups showed no wound width on Day 14 and it indicated a complete wound closure.

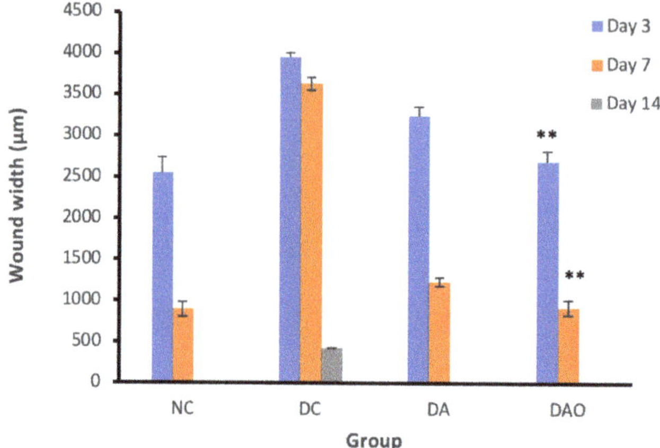

Figure 5. Graph of the wound width parameters. Notation on the graph shows the result of the statistical test ($\alpha < 0.05$). (**) notation shows insignificant differences with NC, but significant differences with DC. ** = $\alpha \leq 0.01$. NC—normal control group; DC—diabetic control group; DA—alginate treatment group; DAO—treatment of alginate–okra fruit extracts. All groups have three variation times.

3.6.2. Neutrophil Measurement

The differences in the number of neutrophils of each group in variations of time were represented in Figure 6. There were significant differences in the number of neutrophils parameters between treatment groups and DC group ($\alpha < 0.05$), but no significant differences between the treatment group and the NC group ($\alpha > 0.05$) were observed. The number of neutrophils increased on Day 3 and decreased on Day 7 and 14 for all groups.

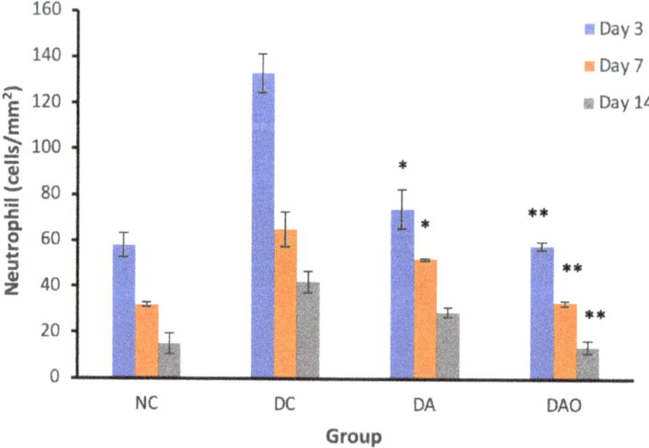

Figure 6. Graph of the number of neutrophils. Notation on the graph shows the result of the statistical test ($\alpha < 0.05$). (*) and (**) notation shows insignificant differences with NC, but significant differences with DC. * = $0.01 \leq \alpha < 0.05$; ** = $\alpha \leq 0.01$. NC—normal control group; DC—diabetic control group; DA—alginate treatment group; DAO—treatment of alginate–okra fruit extracts. All groups have three variation times.

3.6.3. Macrophage Measurement

The difference in the number of macrophages each group in time variations were represented in Figure 7. Similar to the neutrophil parameter, the number of macrophages of DAO group was significant difference to the DC ($\alpha < 0.05$), but no significant difference with NC group ($\alpha > 0.05$). The increase of macrophages occurred up to Day 7 and decreased on Day 14 for all groups.

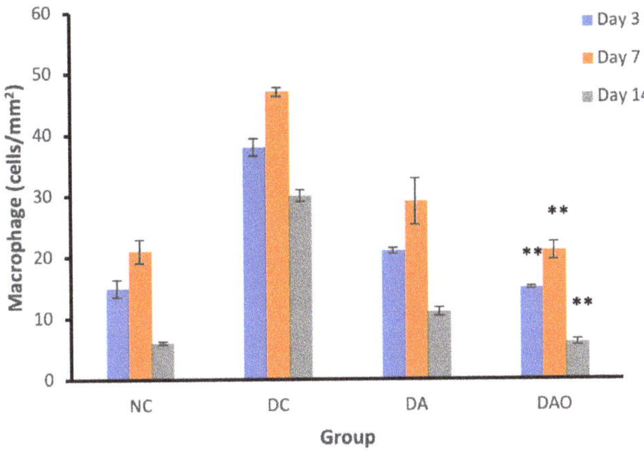

Figure 7. Graph of the number of macrophages. Notation on the graph shows the result of the statistical test ($\alpha < 0.05$). (**) notation shows insignificant differences with NC, but significant differences with DC. ** = $\alpha \leq 0.01$. NC—normal control group; DC—diabetic control group; DA—alginate treatment group; DAO—treatment of alginate–okra fruit extracts. All groups have three variation times.

3.6.4. Fibrocyte and Fibroblast Measurements

The difference in the number of fibrocytes and fibroblasts of each group in time variations are shown in Figure 8a,b. There were significant differences between DAO treatment with DC groups ($\alpha < 0.05$), but no significant difference with NC group ($\alpha > 0.05$). The number of fibrocytes and fibroblasts increased on Day 7 and decreased on Day 14 for all groups.

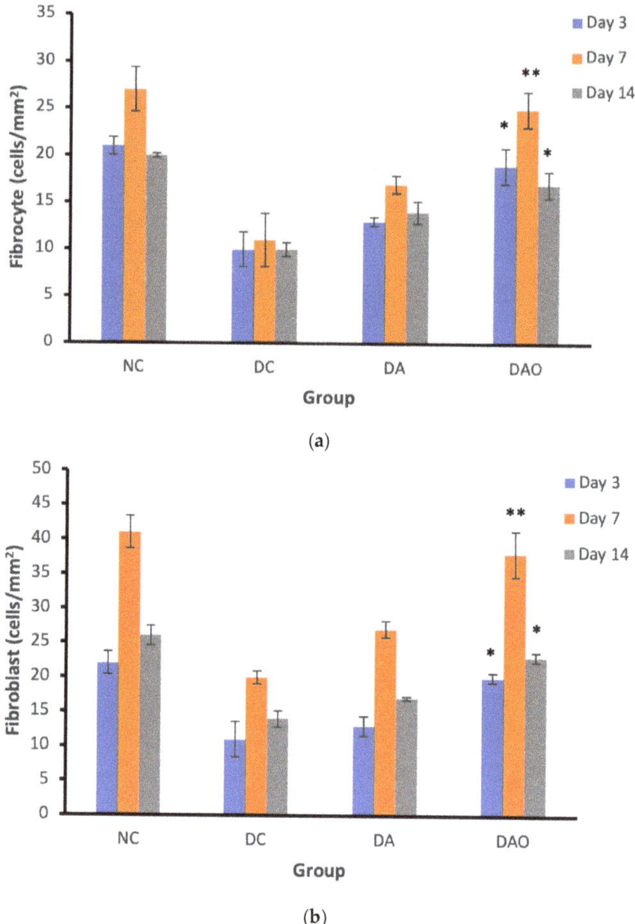

Figure 8. (**a**) Graph of the number of fibrocytes; (**b**) graph of the number of fibroblasts. Notation on the graph showed the result of the statistical test ($\alpha < 0.05$). (*) and (**) notation shows insignificant differences with NC, but significant differences with DC. * = $0.01 \leq \alpha < 0.05$; ** = $\alpha \leq 0.01$. NC—normal control group; DC—diabetic control group; DA—alginate treatment group; DAO—treatment of alginate–okra fruit extracts. All groups have three variation times.

3.6.5. Collagen Densities Measurement

The different collagen densities of each group in time variations are shown in Figure 9. There were significant differences between treatment groups with DC ($\alpha < 0.05$), but no significant difference with NC group ($\alpha > 0.05$) was observed.

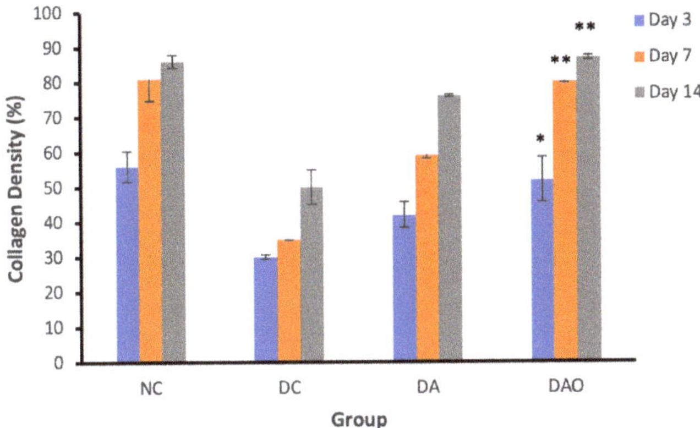

Figure 9. Graph of the collagen densities. Notation on the graph shows the result of the statistical test ($\alpha < 0.05$). (*) and (**) notation shows insignificant differences with NC, but significant differences with DC. * = $0.01 \leq \alpha < 0.05$; ** = $\alpha \leq 0.01$. NC—normal control group; DC—diabetic control group; DA—alginate treatment group; DAO—treatment of alginate–okra fruit extracts. All groups have three variation times.

4. Discussion

The extraction yield of alginate from *S. duplicatum* was higher in comparison with the other seaweed such as *Cystoseira barbata*, *Dictyota caribaea* and *Padina perindusiata* of 9.9, 7.4 and 5.4%, respectively [39,40]. However, the extraction yield was lower than alginate from *C. implexa* and *L. variegate* of 29.15 and 27.57% [41], *S. angustifolium* and *Sargassum* sp. of 40.78 and 44.32%, respectively [42,43]. In general, the content of alginate in various types of brown seaweed varies widely are depending on species and the method for extraction used [44]. The extraction yield and the color of alginate from *S. duplicatum* in this study like to the previous published literature which have pale yellow color and 12% yield [40].

Based on result of characterization of alginate *S. duplicatum*, polymer of alginate *S. duplicatum* is homogeneous because it has polydispersity index is less than two [45]. M_w and M_n of alginate of *S. duplicatum* were lower in comparison to M_w and M_n of *S. turbinarioides* (5.528×10^5 and 3.852×10^5 g/mol) and *S. vulgare* (1.10×10^5 and 1.94×10^5 g/mol) [46,47]. Alginate of *S. duplicatum* has residue of guluronate higher than mannuronate (M/G ratio = 0.91) and also it has η value as 0.68 (less than 1). Hence, it indicated that alginate extracted from *S. duplicatum* is a homopolymer block containing polyguluronate type structure [47]. The M/G ratio of alginate *S. duplicatum* is higher than *S. fluitans* (0.15–0.69), but lower than *S. siliquosum* (0.70–0.94) [48]. Some other *Sargassum* species have M/G ratio about 0.8–1.4 [25].

Quantitatively, okra fruit extracts has a total flavonoid and quercetin content higher from green okra extract of 27.0 mg/100 g and 20.03 mg (QE)/100 g [49,50]. The green okra fruits contain of flavonoid such as a quercetin derivative. The primary flavonoids of okra fruits are up to 70% of the total antioxidant compounds [51]. Total phenolic or total flavonoids in okra extracts play an important role in the ability of antioxidant activity in varying percentages [14]. Antioxidant activity could be classified based on the IC_{50} values, if the values less than 50 µg/mL is very strong/highly active, 51–100 µg/mL is strong/active, 101–250 µg/mL is moderate, 251–500 µg/mL is weak and more than 500 µg/mL is inactive antioxidants [38]. Okra and AOEs have strong inhibition 65.87 and 79.34, respectively in comparison with alginate of 125.31 at the same concentration. Sodium alginate from Cystoseira barbata inhibited of 174 radical scavenging activity of DPPH [12]. It showed that the antioxidant activity of alginates from different species has quite different activities, because antioxidant activity of plant extracts is associated with flavonoid or phenolic compounds in the extract and it depends on the arrangement of structural functional groups of the extracts. In addition, the existence of hydroxyl group, monosaccharide composition, molecular hydrogen bonds, and molecular weight affects the activity of antioxidants [52]. Some structural features of polysaccharides such as molecular weight, monosaccharide composition (block G and M), availability of hydroxyl groups, and molecular hydrogen bonds have an effect on scavenging activity. Alginate that contain block G higher than block M have an increase in antioxidant activity due to diaxial linkage in block G. It causes the obstructed rotation around the glycosidic linkage, so that it can reduce the flexibility of block G. it affects the availability of the hydroxyl group and increased proton donor capability for the free radical scavenger process in alginate molecule [47]. The addition of okra fruit extract from extracts combination on antioxidant activity showed the flavonoid content in okra fruit extract plays an important role in free radical scavenging. The reduction activity of phenolic hydroxyl groups causes flavonoids to contribute hydrogen atoms so that delocalization of phenoxy radical products occurs to protect tissue damage from ROS [53]. In this research, the addition of alginate to the extract combination reduced the IC_{50} value compared to okra fruit extracts. However, alginate can stabilize flavonoids in okra fruit extracts because G and M block residues tend to form diaxial links with intramolecular or extramolecular hydrogen bonds in more stable flavonoid compounds [54].

The administration of lard increase hyperlipidemia and insulin resistance that leads to Type 2 diabetes mellitus which was marked by obesity [30]. Onset of obesity due to the accumulation of excess fat can lead to chronic diseases and complications such as diabetes mellitus and cardiovascular disease [55]. STZ administration is expected to increase the condition of hyperglycemia (increase blood-glucose levels) and created diabetic-like syndrom [29]. The destruction of the pancreatic cells were carried out by free radical of STZ which was toxic to the insulin sensitive tissues, so that the secretion of insulin hormone was decrease [56]. Topical treatment using ointment was expected to lowering blood-glucose levels that lead to the wound healing on diabetic mice. Removal of the stratum corneum layer in the skin layer during wound formation increase skin permeability thereby allowing the release of active compounds from topical formulations in the skin layer to enter the systemic circulation. In addition, active compounds pass penetration through the transappendageal route (sweat glands, hair follicles and sebaceous glands which include open channels on the outside surface of the skin) [57,58]. Tan et al. [59] revealed that the topical application of Vicenin-2 film which is a type of flavonoid-derived glycosides from various natural plants can reduce blood-glucose levels in diabetic mice. Hence, that, lowering blood-glucose levels on diabetic mice can be influenced by topical treatment of open wounds. The combination of extracts could reduce the condition of diabetes which is demonstrated in Table 2. The value observed of DAO group at day 14 was almost the same of the control group (115.0 ± 19.1). it is indicating the potentiality of okra fruit extract in reducing blood-glucose level on diabetic mice. Antioxidant properties can neutralize free radical damage so that blood-glucose levels decreases. Okra fruit extract can control blood-glucose levels, as an effective treatment for diabetic ulcers [60].

No significant differences between DAO group (2693.0 ± 12.2; 917.0 ± 91.3; 0 ± 0 μm) and normal group (NC) (2548.0 ± 187.2; 894.0 ± 87.6; 0 ± 0 μm) on Days 3, 7 and 14 (Table 3). This indicates that the topical administration of alginate and okra fruit extracts combination can increase re-epithelization of the wound area in comparison to the control group. Diabetic group (DC) showed that still has wound width on Day 14 (424.0 ± 5.2 μm). Otherwise, NC, DA and DAO groups showed no wound width on Day 14 (0 ± 0 μm) and it indicated a complete wound closure.

Table 3. Counts considering the mean and standard deviation of wound width.

Group Treatment	Wound Width (μm)		
	Day 3	Day 7	Day 14
Normal (NC)	2548.0 ± 187.2	894.0 ± 87.6	0 ± 0
Diabetic (DC)	3956.0 ± 52.2	3631.0 ± 74.8	424.0 ± 5.2
Alginate ointment (DA)	3240.0 ± 117.5	1229.0 ± 49.9	0 ± 0
Alginate–okra ointment (DAO)	2693.0 ± 12.2 **	917.0 ± 91.3 **	0 ± 0

Letter notation on the table shows the result of the statistical test ($\alpha < 0.05$). (**) notation shows insignificant differences with NC, but significant differences with DC. ** = $\alpha \leq 0.01$.

The number of neutrophils increased on Day 3 and decreased up to Day 14 for all groups, due to migration of neutrophil achieved a maximum between Day 1 and 2. The plateaued level on Day 3 and decreases on Day 5 [61]. Neutrophil secretions in diabetic conditions tend to be high compared to the normal conditions due to high ROS/RNS levels, resulting in increased inflammation and tissue damage [62]. This is evidence by the higher number of neutrophils on Day 3 up to Day 14 in the diabetic control group (133.0 ± 8.2; 65.0 ± 7.5; 42.0 ± 4.7 cells/mm^2) compared to normal control group (58.0 ± 5.4; 32.0 ± 0.9; 15.0 ± 4.5 cells/mm^2) (Table 4).

Table 4. Counts considering the mean and standard deviation of neutrophils.

Group Treatment	Neutrophils (cells/mm^2)		
	Day 3	Day 7	Day 14
Normal (NC)	58.0 ± 5.4	32.0 ± 0.9	15.0 ± 4.5
Diabetic (DC)	133.0 ± 8.2	65.0 ± 7.5	42.0 ± 4.7
Alginate ointment (DA)	74.0 ± 8.5 *	52.0 ± 0.5 *	29.0 ± 2.1
Alginate–okra ointment (DAO)	58.0 ± 1.9 **	33.0 ± 1.2 **	14.0 ± 2.6 **

Letter notation on the table shows the result of the statistical test ($\alpha < 0.05$). (*) and (**) notation shows insignificant differences with NC, but significant differences with DC. * = $0.01 \leq \alpha < 0.05$; ** = $\alpha \leq 0.01$.

Number of macrophages increase up to Day 7 and decreased on Day 14 for all groups, because on the Day 7 the wound condition undergoes inflammatory and proliferation processes. For example in the DAO group on Day 3 up to Day 7 number of macrophages increased (15.0 ± 0.2; 21.0 ± 1.4 cells/mm^2) and decreased on Day 14 (6.0 ± 0.5 cells/mm^2) (Table 5). Neutrophils were secreting cytokines and granule proteins, which modulating monocyte or macrophage extravasation, phagocytosis and ROS production. In diabetics, the condition of hyperglycemia increases the activity of macrophages, thereby increasing the levels of ROS/RNS which causes a prolonged inflammatory phase [63] In addition, hyperglycemia and oxidative stress conditions can lead to polarization and modulating macrophage dysregulation, because change epigenic code which is inhibit the wound-healing process [64]. This is evidence by the higher number of macrophages on Day 3 up to Day 14 in the diabetic control group (38.0 ± 1.4; 47.0 ± 0.7; 30.0 ± 0.9 cells/mm^2) compared to normal control group (15.0 ± 1.4; 21.0 ± 1.9; 6.0 ± 0.2 cells/mm^2).

Table 5. Counts considering the mean and standard deviation of macrophages.

Group Treatment	Macrophages (cells/mm^2)		
	Day 3	Day 7	Day 14
Normal (NC)	15.0 ± 1.4	21.0 ± 1.9	6.0 ± 0.2
Diabetic (DC)	38.0 ± 1.4	47.0 ± 0.7	30.0 ± 0.9
Alginate ointment (DA)	21.0 ± 0.5	29.0 ± 3.8	11.0 ± 0.7
Alginate–okra ointment (DAO)	15.0 ± 0.2 **	21.0 ± 1.4 **	6.0 ± 0.5 **

Letter notation on the table shows the result of the statistical test ($\alpha < 0.05$). (**) notation shows insignificant differences with NC, but significant differences with DC. ** = $\alpha \leq 0.01$.

The number of fibrocytes and fibroblasts increased on Day 7 and decreased on Day 14 for all groups, because the wound conditions have peak proliferation phases for re-epithelization and angiogenesis processes [65]. For example in the DAO group on Day 3 up to Day 7 number of fibrocyte increased (19.0 ± 1.9; 25.0 ± 1.9 cells/mm^2) and decreased on Day 14 (17.0 ± 1.4 cells/mm^2) (Table 6). The same thing happened to the number of fibroblasts in the DAO group on Day 3 up to Day 7 increased (20.0 ± 0.7; 38.0 ± 3.3 cells/mm^2) and decreased on Day 14 (23.0 ± 0.7 cells/mm^2) (Table 7). Macrophages perform cytokine and growth factor secretions in the wound area to induce fibroblast migration and proliferation, production of granulation tissues, transient extracellular matrix and angiogenesis in the healing process. Conditions of high blood-glucose levels lead to migration and fibrocyte proliferation disorders, fibroblasts and keratinocytes, resulting in decreased secretion of cytokine, growth factor and extracellular matrix [63]. This is evidence by the lower number of fibrocytes on Day 3 up to Day 14 in the diabetic control group (10.0 ± 1.9; 11.0 ± 2.8; 10.0 ± 0.7 cells/mm^2) compared to normal control group (21.0 ± 0.9; 27.0 ± 2.3; 20.0 ± 0.2 cells/mm^2). The same thing happened to the number of fibroblasts on Day 3 up to Day 14 in the diabetic control group (11.0 ±2.6; 20.0 ± 0.9; 14.0 ± 1.2 cells/mm^2) compared to normal control group (22.0 ± 1.6; 41.0 ± 2.4; 26.0 ± 1.4 cells/mm^2).

Table 6. Counts considering the mean and standard deviation of fibrocytes.

Group Treatment	Fibrocytes (cells/mm^2)		
	Day 3	**Day 7**	**Day 14**
Normal (NC)	21.0 ± 0.9	27.0 ± 2.3	20.0 ± 0.2
Diabetic (DC)	10.0 ± 1.9	11.0 ± 2.8	10.0 ± 0.7
Alginate ointment (DA)	13.0 ± 0.5	17.0 ± 0.9	14.0 ± 1.2
Alginate–okra ointment (DAO)	19.0 ± 1.9 *	25.0 ± 1.9 **	17.0 ± 1.4 *

Letter notation on the table shows the result of the statistical test ($\alpha < 0.05$). (*) and (**) notation shows insignificant differences with NC, but significant differences with DC. * = $0.01 \leq \alpha < 0.05$; ** = $\alpha \leq 0.01$.

Table 7. Counts considering the mean and standard deviation of fibroblasts.

Group Treatment	Fibroblasts (cells/mm^2)		
	Day 3	**Day 7**	**Day 14**
Normal (NC)	22.0 ± 1.6	41.0 ± 2.4	26.0 ± 1.4
Diabetic (DC)	11.0 ±2.6	20.0 ± 0.9	14.0 ± 1.2
Alginate ointment (DA)	13.0 ± 1.4	27.0 ± 1.2	17.0 ± 0.2
Alginate–okra ointment (DAO)	20.0 ± 0.7 *	38.0 ± 3.3 **	23.0 ± 0.7 *

Letter notation on the table shows the result of the statistical test ($\alpha < 0.05$). (*) and (**) notation shows insignificant differences with NC, but significant differences with DC. * = $0.01 \leq \alpha < 0.05$; ** = $\alpha \leq 0.01$.

Fibroblast proliferation induces collagen synthesis and macromolecular matrix for structural formation of connective tissues. Diabetes conditions increase the activity of macrophages thereby increasing the production of ROS/RNS and lowering collagen synthesis [66]. This is evidence by the lower collagen densities on Day 3 up to Day 14 in the diabetic control group (30.0 ± 0.6; 35.0 ± 0.1; 50.0 ± 4.9%) compared to normal control group (56.0 ± 4.3; 81.0 ± 6.4; 86.0 ± 1.9%) (Table 8).

Table 8. Counts considering the mean and standard deviation of collagen densities.

Group Treatment	Collagen Densities (%)		
	Day 3	**Day 7**	**Day 14**
Normal (NC)	56.0 ± 4.3	81.0 ± 6.4	86.0 ± 1.9
Diabetic (DC)	30.0 ± 0.6	35.0 ± 0.1	50.0 ± 4.9
Alginate ointment (DA)	42.0 ± 3.8	59.0 ± 0.8	76.0 ± 0.3
Alginate–okra ointment (DAO)	52.0 ± 6.4 *	80.0 ± 0.3 **	87.0 ± 0.6 **

Letter notation on the table shows the result of the statistical test ($\alpha < 0.05$). (*) and (**) notation shows insignificant differences with NC, but significant differences with DC. * = $0.01 \leq \alpha < 0.05$; ** = $\alpha \leq 0.01$.

The topical administration of alginate can reduce the wounds width and increase cytokines and growth factor. This is because alginate has a hydroxyl group that plays a role in the properties of hydrophilicity, so that alginate can absorb moderate or heavy exudate liquid to reach 15 to 20 times in dry or gel form, so it is called a hemostatic agent [67]. Absorption of exudate on the wound area was occur due to the exchange of ions between exudate and alginate that keeps the moist condition of the wound physiologically and minimizing bacterial infections. In addition, it allows enough oxygen exchange to accelerate the formation of tissue granulation and re-epithelization [68]. The proportion of M/G residue affects alginate's ability as absorber. Higher G content is not firmly tied to molecules so it will increase the process of ion exchange between alginate and exudate wounds [69]. Based on literature, the wound healing activity of alginate topically on diabetic open wounds form complete epithelization on Day 14 [70].

The topical administration of okra fruit extract can reduce blood-glucose levels and improve wound-healing process, because it has strong antioxidant properties due to the role of the 3-hydroxyl group on the C ring of the flavonoids compounds to capture free radicals due to glucose oxidation (hyperglycemia). Decreasing ROS/RNS can reduce hyperglycemia and increase glucose metabolism, thereby lowering oxidative stress and insulin resistance which increase wound healing [71]. The ointment formula contains flavonoids could be released into wounds and then to the bloodstream so as to induce secretion of cytokine, growth factor and increase insulin production or insulin sensitivity of somatic cells. Flavonoids also cause re-formation of antioxidants and serum lipid profiles in diabetic rats [59]. The wound healing activity of the okra fruit extract topically on open wounds effectively forms the complete epithelization on 15 day [72] and achieves a complete epithelization on 16 day [73].

5. Conclusions

Extraction and characterization of alginate from *S. duplicatum* and okra fruit extracts were carried out to determine the potentiality of them as an absorber of the wound and their antioxidant activity. The novelty of this research lies in the healing activity of open wounds from a combination of alginate and okra fruit extracts. Topical formulations are more effective when alginates are combined with okra fruit extracts because of the alginate can stabilize flavonoids, so it increased wound absorber and their antioxidant properties. The results show an improvement of the re-epithelization of the wound area, the recruitment of neutrophil, macrophages, fibrocytes, fibroblasts and increased collagen densities as parameters that plays an important role in the wound-healing process.

Further research and a more detailed analysis of the characterization of alginate structures to determine the role of alginate as an antioxidant agent in the combination of topical formulations for open wounds healing are needed. Moreover, variations of doses need to be done for further research to determine the optimum dose of the extract combination in open wound healing of diabetic mice. Hence, it would help to provide more insight of the mechanism of action this combination extracts on the process of diabetic wound healing.

Author Contributions: This paper is a part of master's degree thesis. Conceptualization, P.P. and D.W.; methodology, Z.N.I. and S.A.H.; software, Z.N.I.; validation, D.W., A.P. and S.A.H.; formal analysis, Z.N.I., A.P. and M.V.; investigation, Z.N.I.; resources, Z.N.I., P.A.C.W. and M.A.A.; data curation, Z.N.I.; writing—original draft preparation, Z.N.I.; writing—review and editing, P.P., K.A. and D.M.; visualization, Z.N.I.; supervision, P.P. and D.W.; project administration, Z.N.I., P.A.C.W., S.A.H.; funding acquisition, P.P. All authors have read and agreed to the published version of the manuscript.

Funding: This research was funded by Innovation and Research Center, Airlangga University, Grant Number 1408/UN3/2019 and The APC was funded by Innovation and Research Center, Airlangga University. A.P. thanks the Austrian Science Fund (FWF) for the funding through the Erwin Schrödinger Fellowship, grant agreement: J4014-N34.

Acknowledgments: The author would like to thank Innovation and Research Center, Airlangga University for supporting this research through Mandat Research Grant, Airlangga University FY 2019 and Leonardo Gomez from the CNAP of the University of York for access to the SEC-MALLS system.

Conflicts of Interest: The authors declare no conflict of interest.

References

1. Das, S.; Majid, M.; Baker, A.B. Syndecan-4 Enhances Pdgf-Bb activity in diabetic wound healing. *Acta Biomater.* **2016**, *15*, 56–65. [CrossRef]
2. Houstis, N.; Rosen, E.D.; Lander, E.S. Reactive oxygen species have a causal role in multiple forms of insulin resistance. *Nature* **2006**, *440*, 944–948. [CrossRef] [PubMed]
3. Gary, S.R.; Woo, K.Y. The biology of chronic foot ulcers in persons with diabetes. *Diabetes/Metab. Res. Rev.* **2008**, *24*, 25–30. [CrossRef]
4. American Diabetes Association. Classification and diagnosis of diabetes mellitus. *Diabetes Care* **2015**, *38*, 8–16. [CrossRef]
5. Gilgun-Sherki, Y.; Melamed, E.; Offen, D. Oxidative stress induced-neurodegenerative diseases: The need for antioxidants that penetrate the blood brain barrier. *Neuropharmacology* **2001**, *40*, 959–975. [CrossRef]
6. Prompers, L.; Schaper, N.; Apelqvist, J.; Edmonds, M.; Jude, E.; Mauricio, D. Prediction of outcome in individuals with diabetic foot ulcers: Focus on the differences between individuals with and without peripheral arterial disease. *Eurodiale Study Diabetol.* **2008**, *51*, 747–755. [CrossRef]
7. Babitha, S.; Rachita, L.; Karthikeyan, K.; Shoba, E.; Janani, I.; Poornima, B. Electrospun protein nanofibers in healthcare: A review. *Int. J. Pharm.* **2017**, *523*, 52–90. [CrossRef]
8. Ode, I. The content of Alginate seaweed of *Sargassum crassifolium* from the coastal waters of Hutumuri village Ambon City. *J. Sci. Agribus. Fish.* **2014**, *6*, 3.
9. Sezer, A.D.; Cevher, E. Biopolymers as wound healing materials: Challenges and new strategies. In *Biomaterials Applications for Nanomedicine*; Pignatello, R., Ed.; InTech: Rijeka, Croatia, 2011; pp. 383–414.
10. Yu, W.; Jiang, Y.Y.; Sun, T.W.; Qi, C.; Zhao, H.; Chen, F.; Shi, Z.; Zhu, Y.J.; Chen, D.; He, Y. Design of a novel wound dressing consisting of alginate hydrogel and simvastatin-incorporated mesoporous hydroxyapatite microspheres for cutaneous wound healing. *R. Soc. Chem.* **2016**, *6*, 104375–104387. [CrossRef]
11. Nazeri, S.; Ardakani, E.M.; Babavalian, H.; Latifi, A.M. Evaluation of Effectiveness of Honey-Based Alginate Hyrogel on Wound Healing in a Mouse Model of Rat. *J. Appl. Biotechnol. Rep.* **2015**, *2*, 293–297.
12. Ahmed, M.M.; Jahangir, M.A.; Saleem, M.A.; Kazmi, I.; Bhavani, P.D.; Muheem, A. Formulation and Evaluation of Fucidin Topical Gel Containing Wound Healing Modifiers. *Am. J. Pharmtech Res.* **2015**, *5*, 232–242.
13. Hegge, A.B.; Andersen, T.; Melvik, J.E.; Bruzell, E.; Kristensen, S.; Tønnesen, H.H. Formulation and bacterial phototoxicity of curcumin loaded alginate foams for wound treatment applications: Studies on curcumin and curcuminoides XLII. *J. Pharm. Sci.* **2011**, *100*, 174–185. [CrossRef]
14. Suzuki, Y.; Nishimura, Y.; Tanihara, M.; Suzuki, K.; Kitahara, A.K.; Yamawaki, Y.; Nakamura, T.; Shimizu, Y.; Kakimaru, Y. Development of alginate gel dressing. *J. Artif. Organs.* **1998**, *1*, 28–32. [CrossRef]
15. Dyson, M.; Young, S.R.; Pendte, L.; Webster, D.F.; Lang, S. Comparison of the effects of moist and dry conditions on dermal repair. *J. Investig. Dermatol.* **1988**, *91*, 434–439. [CrossRef]
16. Sellimi, S.; Younes, I.; Ayed, H.B.; Maalej, H.; Montero, V.; Rinaudo, M.; Dahia, M.; Mechichi, T.; Hajji, M.; Nasri, M. Structural, physicochemical and antioxidant properties of sodium alginate isolated from a Tunisian brown seaweed. *Int. J. Biol. Macromol.* **2015**, *72*, 1358–1367. [CrossRef]
17. Szekalska, M.; Puciłowska, A.; Szymańska, E.; Ciosek, P.; Winnicka, K. Alginate: Current Use and Future Perspectives in Pharmaceutical and biomedical applications. *Int. J. Polym. Sci.* **2016**, *8*, 1–17. [CrossRef]
18. Liao, H.; Liu, H.; Yuan, K. A New Flavonol glycoside from the *Abelmoschus esculentus* Linn. *Pharmacogn. Mag.* **2012**, *8*, 12–15. [CrossRef]
19. Sengkhamparn, N.; Verhoef, R.; Schols, H.; Sajjaanantakul, T.; Voragen, A.G. Characterisation of cell wall polysaccharides from okra (*Abelmoschus esculentus* (L.) Moench). *Carbohydr. Res.* **2009**, *344*, 1824–1832. [CrossRef]
20. Sabitha, V.; Ramachandran, S.; Naveen, K.R.; Panneerselvam, K. Antidiabetic and antihyperlidemic activity of *Abelmoschus esculentus* (L.) Moench. in streptozotocin-induced diabetic rats. *J. Pharm. Bioall. Sci.* **2011**, *3*, 397–402. [CrossRef]
21. Gouma, E.; Simos, Y.; Verginadis, I.; Batistatou, S.; Karkabounas, S.; Evangelou, A.; Ragos, V.; Peschos, D. Healing effects of Quercetin on full thickness epidermal thermal injury in Wistar rats. *Int. J. Phytomedicine* **2016**, *8*, 277–281.
22. Rasyid, A. Algae coklat (*Phaeophyta*) as source of alginate: Oceanography LIPI. *Oseana* **2003**, *28*, 33–38.

23. Vastano, M.; Pellis, A.; Botelho-Machado, C.; Simister, S.; McQueen-Mason, S.J.; Farmer, T.J.; Gomez, L.D. Sustainable Galactarate-Based Polymers: Multi-Enzymatic Production of Pectin-Derived Polyesters. *Macromol. Rapid Commun.* **2019**, *40*, 1900361. [CrossRef] [PubMed]
24. Rinaudo, M. Main properties and current applications of some polysaccharides as biomaterials. *Polym. Int.* **2008**, *3*, 397–430. [CrossRef]
25. Llanes, F.; Sauriol, F.; Morin, F.G.; Perlin, A.S. An examination of sodium alginate from *Sargassum* by NMR spectroscopy. *Can. J. Chem.* **1997**, *75*, 585–590. [CrossRef]
26. Devi, N.; Kakati, D.K. Smart porous microparticles based on gelatin/sodium alginate polyelectrolyte complex. *J. Food Eng.* **2006**, *117*, 193–204. [CrossRef]
27. Ahiakpa, J.K.; Amoatey, H.M.; Amenorpe, G.; Apatey, J.; Ayeh, E.A.; Quartey, E.K.; Agbemavor, W.S. Mucilage content of 21 accessions of okra (*Abelmoschus* spp L.). *Sci. Agric.* **2014**, *2*, 96–101.
28. Prieto, J.M. *Procedure: Preparation of DPPH Radical and Antioxidant Scavenging Assay, Dr. Prieto's Dpph Microplate Protocol*; School of Pharmacy and Biomolecular Science, Liverpool John Moores University: Liverpool, UK, 2012; pp. 1–3.
29. Raisi-Nafchi, M.; Kavoosi, G.; Nasiri, S.M. Physico-mechanical and antioxidant properties of carboxymethylcellulose and alginate dispersions and essential oil based films for use as food packaging materials. *Acad. J. Food Res.* **2016**, *4*, 001–010. [CrossRef]
30. Novelli, M.; Bonamasa, B.; Masini, M.; Funel, N.; Canistro, D.; Martano, M.; Soleti, A. Persistent correction of hyperglycemia in streptozotocin-nicotinamide-induced diabetic mice by a non-conventional radical scavenger. *Naunyn Schmied. Arch. Pharmacol.* **2010**, *382*, 127–137. [CrossRef] [PubMed]
31. Federer, W.T. *Experimental Design Theory and Application*; Oxford & IBH: Calcutta, India, 1967.
32. Rinaudo, M. Seaweed polysaccharaides. In *Comprehensive Glycoscience: From Chemistry to System Biology*; Kalmering, J.P., Ed.; Elsevier Science: Amsterdam, The Netherlands, 2007; Volume 2, pp. 691–735.
33. Soares, S.; Cammino, G.; Levchick, S. Comparative study of the thermal decomposition of pure cellulose and pulp paper. *Polym. Degrad. Stab.* **1995**, *49*, 275–283. [CrossRef]
34. Patel, N.; Lalwani, D.; Gollmer, S.; Injeti, E.; Sari, Y.; Nesamony, J. Development and evaluation of a calcium alginate based oral ceftriaxone sodium formulation. *Prog. Biomater.* **2016**, *5*, 117–133. [CrossRef]
35. Grasdalen, H.; Larsen, B.; Smidsrod, O. A P.M.R. study of the composition and sequence of urinate residues in alginates. *Carbohydr. Res.* **1979**, *68*, 23–31. [CrossRef]
36. Cui, S.W. *Food Carbohydrates: Chemistry, Physical Properties, and Application*, 1st ed.; CRC Press: Boca Raton, FL, USA, 2005.
37. Jensen, H.M.; Larsen, F.H.; Engelsen, S.B. Characterization of alginates by Nuclear Magnetic Resonance (NMR) and vibrational spectroscopy (IR, NIR, Raman) in combination with chemometrics. *Nat. Prod. Mar. Algae* **2015**, *1308*, 347–363.
38. Jun, M.; Fu, H.Y.; Hong, J.; Wan, X.; Yang, C.S. Comparison of antioxidant activities of isoflavones from kadzu root (*Puerari lobata ohwi*). *J. Food Sci.* **2003**, *68*, 2117–2122. [CrossRef]
39. Sellimi, S.; Maalej, H.; Rekik, D.M.; Benslima, A.; Ksouda, G.; Hamdi, M.; Sahnoun, Z.; Li, S.; Nasri, M.; Hajji, M. Antioxidant, antibacterial and in vivo wound healing properties of laminaran purified from Cystoseira barbata seaweed. *Int. J. Biol. Macromol.* **2018**, *119*, 633–644. [CrossRef]
40. Nishinari, K.; Doi, E. *Food Hydrocolloids: Structure, Properties, and Functions*; Plenum Press: New York, NY, USA, 1993.
41. Viswanathan, S.; Nallamuthu, T. Extraction of sodium alginate from selected seaweeds and their physiochemical and biochemical properties. *Int. J. Innov. Res. Sci. Eng. Technol.* **2014**, *3*, 2319–8753.
42. Gholamipoor, S.; Nikpour-Ghanavati, Y.; Oromiehie, A.R.; Mohammadi, M. Extraction and characterization of alginate from Sargassum angustifolium collected from Northern Coasts of Persian Gulf, Bushehr. *Int. Symp. Adv. Sci. Technol.* **2013**, *1*, 1 5.
43. Helmiyati, A.M. Characterization and properties of sodium alginate from brown algae used as an ecofriendly superabsorbent. *IOP Conf. Ser. Mater. Sci. Eng.* **2017**, *188*, 12019. [CrossRef]
44. Larsen, B.; Salem, D.M.S.A.; Sallan, M.A.E.; Mishrikey, M.M.; Beltagy, A.I. Characterization of the alginates from algae harvested at the Egyptian Red Seacoast. *Carbohydr. Res.* **2003**, *338*, 2325–2336. [CrossRef]
45. Atkins, P.; Paula, J.D. *Physical Chemistry*, 9th ed.; University Press: Oxford, UK, 2010.

46. Fenoradosoa, T.A.; Ali, G.; Delattre, C.; Laroche, C.; Petit, E.; Wadouachi, A.; Michaud, P. Extraction and characterization of an alginate from the brown seaweed Sargassum turbinarioides Grunow. *Environ. Boil. Fishes* **2009**, *22*, 131–137. [CrossRef]
47. Sari-Chmayssem, N.; Taha, S.; Mawlawi, H.; Guégan, J.; Jeftić, J.; Benvegnu, T. Extracted and depolymerized alginates from brown algae Sargassum vulgare of Lebanese origin: Chemical, rheological, and antioxidant properties. *J. Appl. Phycol.* **2016**, *28*, 1915–1929. [CrossRef]
48. Davis, T.A.; Llanes, F.; Volesky, B.; Diaz-Pulido, G.; McCook, L.; Mucci, A. ^{1}H-NMR study of Na alginates extracted from *Sargassum* spp. in relation to metal biosorption. *Appl. Biochem. Biotechnol.* **2003**, *110*, 75–90. [CrossRef]
49. Anjani, P.P.; Damayanthi, E.; Handharyani, E. Antidiabetic potential of purple okra (Abelmoschus esculentus L.) extract in streptozotocin-induced diabetic rats. *IOP Conf. Ser. Earth Environ. Sci.* **2018**, *196*, 012038. [CrossRef]
50. Savova-Tsanova, S.; Ribarova, F.; Petkov, V. Quercetin content and ratios to total flavonols and total flavonoids in Bulgarian fruits and vegetables. *Bulg. Chem. Commun.* **2018**, *50*, 69–73.
51. Shui, G.; Peng, L.L. An improved method for the analysis of major antioxidants of *Hibiscus esculentus* Linn. *J. Chromatogr.* **2004**, *1048*, 17–24. [CrossRef]
52. Lo, T.C.; Chang, C.A.; Chiu, K.; Tsay, P.; Jen, J. Correlation evaluation of antioxidant properties on the monosaccharide components and glycosyl linkages of polysaccharide with different measuring methods. *Carbohydr. Polym.* **2011**, *86*, 320–327. [CrossRef]
53. Verma, A.K.; Singh, H.; Satyanarayana, M. Flavone-based novel antidiabetic and antidyslipidemic agents. *J. Med. Chem.* **2012**, *55*, 4551–4567. [CrossRef]
54. Wei, C.; Yang, X.; Wang, D.; Fang, F.; Lai, J.; Wang, F.; Wu, T. Fatty acid composition and evaluation on antioxidation activities of okra seed oil under ultrasonic wave extraction. *Cereals Oils Assoc.* **2016**, *31*, 89–93.
55. Husen, S.A.; Wahyuningsih, S.P.A.; Ansori, A.N.M.; Hayaza, S.; Susilo, R.J.K.; Darmanto, W.; Winarni, D. The effect of okra (*Abelmoschus esculentus* Moench) pods extract on malondialdehyde and cholesterol level in STZ-induced diabetic mice. *Ecol. Environ. Conserv.* **2019**, *25*, 50–56.
56. Husen, S.A.; Khaleyla, F.; Ansori, A.N.M.; Susilo, R.J.K.; Winarni, D. Antioxidant activity assay of alpha-mangostin for amelioration of kidney structure and function in diabetic mice. *Adv. Soc. Sci. Educ. Humanit. Res.* **2018**, *98*, 84–88.
57. Andrews, S.N.; Jeong, E.; Prausnitz, M.R. Transdermal delivery of molecules is limited by full epidermis, not just stratum corneum. *Pharm. Res.* **2013**, *30*, 1099–1109. [CrossRef]
58. Mathes, S.H.; Ruffner, H.; Graf-Hausner, U. The use of skin models in drug development. *Adv. Drug Deliv. Rev.* **2014**, *69*, 81–102. [CrossRef] [PubMed]
59. Tan, W.S.; Arulselvan, P.; Ng, S.; Taib, C.N.M.; Sarian, M.N.; Fakurazi, S. Improvement of diabetic wound healing by topical application of Vicenin-2 hydrocolloid film on Sprague Dawley rats. *Complementary Altern. Med.* **2019**, *19*, 20. [CrossRef]
60. Jain, N. A review on Abelmoschus esculentus. *Pharmacacia* **2012**, *1*, 1–8.
61. Kim, M.H.; Liu, W.; Borjesson, D.L.; Curry, F.R.; Miller, L.S.; Cheung, A.L.; Liu, F.T.; Isseroff, R.R.; Simon, S.I. Dynamics of neutrophil infiltration during cutaneous wound healing and infection using fluorescence imaging. *J. Investig. Dermatol.* **2008**, *128*, 1812–1820. [CrossRef] [PubMed]
62. Wong, S.L.; Demers, M.; Martinod, K.; Gallant, M.; Wang, Y.; Goldfine, A.B. Diabetes primes neutrophils to undergo NETosis, which impairs wound healing. *Nat. Med.* **2015**, *21*, 815–819. [CrossRef]
63. Patel, S.; Srivastava, S.; Singh, M.R.; Singh, D. Mechanistic insight into diabetic wounds: Pathogenesis, molecular targets and treatment strategies to pace wound healing. *Biomed. Pharmacother.* **2019**, *112*, 1086–1115. [CrossRef]
64. Mallik, S.B.; Jayashree, B.S.; Shenoy, R.R. Epigenetic modulation of macrophage polarization-perspectives in diabetic wounds. *J. Diabetes Complicat.* **2018**, *32*, 524–530. [CrossRef]
65. Smith, J.L.; Sheffield, L.G. Production and regulation of leptin in bovine mammary epithelial cells. *Domest. Anim. Endocrinol.* **2002**, *22*, 145–154. [CrossRef]
66. Heublein, H.; Bader, A.; Giri, S. Preclinical and clinical evidence for stem cell therapies as treatment for diabetic wounds. *Drug Discov. Today* **2015**, *20*, 703–717. [CrossRef]
67. Bello, Y.M.; Philips, T.J. Recent advance in wound healing. *JAMA* **2000**, *283*, 716–718. [CrossRef]
68. Aderibigbe, B.A.; Buyana, B. Alginate in wound dressing. *Pharmaceutics* **2018**, *10*, 42. [CrossRef] [PubMed]

69. Uzun, M. A Review of wound management materials. *J. Text. Eng. Fash. Technol.* **2018**, *4*, 53–59. [CrossRef]
70. Wang, T.; Gu, Q.; Zhao, J.; Mei, J.; Shao, M.; Pan, Y.; Zhang, J.; Wu, H.; Zhang, Z.; Liu, F. Calcium alginate enhances wound healing by up-regulating the ratio of collagen types I/III in diabetic rats. *Int. J. Clin. Exp. Pathol.* **2015**, *8*, 6636–6645.
71. Wahyuningsih, S.P.A.; Pramudya, M.; Putri, I.P.; Winarni, D.; Savira, N.I.I.; Darmanto, W. Crude polysaccharides from okra pods (*Abelmoschus esculentus*) grown in Indonesia enhance the immune response due to bacterial infection. *Adv. Pharmacol. Sci.* **2018**, *1*, 450–538. [CrossRef] [PubMed]
72. Capapas, C.A.; Reas, D.D. Wound Healing Activity of the Extract from the Fresh Fruits of Okra, *Abelmoschus esculentus* (Linn.,1753) on Rabbits. Ph.D. Thesis, University of San Carlos-Josef Baumgartner Learning Resource Center, Science and Technology Section, Cebu, Philippines, March 2012.
73. Farooqui, M.B.; Khasim, S.M.; Rafiq, M.; Khan, A.A. Evaluation of wound healing activity of *Abelmoschus esculentus* (linn) in albino wistar rats. *Eur. J. Pharm. Med. Res.* **2018**, *5*, 508–511.

Article

Wound Healing and Antioxidant Evaluations of Alginate from *Sargassum ilicifolium* and Mangosteen Rind Combination Extracts on Diabetic Mice Model

Pugar Arga Cristina Wulandari [1], Zulfa Nailul Ilmi [1], Saikhu Akhmad Husen [2], Dwi Winarni [2], Mochammad Amin Alamsjah [3], Khalijah Awang [4], Marco Vastano [5], Alessandro Pellis [5,6], Duncan MacQuarrie [5] and Pratiwi Pudjiastuti [1,*]

1 Department of Chemistry, Faculty of Science and Technology, Airlangga University, Surabaya 60115, Indonesia; cristinapugar@gmail.com (P.A.C.W.); zulfanailulilmi@gmail.com (Z.N.I.)
2 Department of Biology, Faculty of Science and Technology, Airlangga University, Surabaya 60115, Indonesia; saikhu-a-h@fst.unair.ac.id (S.A.H.); dwi-w@fst.unair.ac.id (D.W.)
3 Department of Marine, Faculty of Fisheries and Marine, Airlangga University, Surabaya 60115, Indonesia; alamsjah@fpk.unair.ac.id
4 Department of Chemistry, Faculty of Science, University of Malaya, Kuala Lumpur 50603, Malaysia; khalijah@um.edu.my
5 Department of Chemistry, University of York, Heslington, York YO10 5DD, UK; marco.vastano@york.ac.uk (M.V.); alessandro.pellis@boku.ac.at (A.P.); duncan.macquarrie@york.ac.uk (D.M.)
6 Department for Agrobiotechnology, IFA-Tulln, Institute for Environmental Biotechnology, University of Natural Resources and Life Sciences, Konrad Lorenz Strasse 20, Tulln an der Donau, 3430 Vienna, Austria
* Correspondence: pratiwi-p@fst.unair.ac.id; Tel.: +62-856-3390-952

Citation: Wulandari, P.A.C.; Ilmi, Z.N.; Husen, S.A.; Winarni, D.; Alamsjah, M.A.; Awang, K.; Vastano, M.; Pellis, A.; MacQuarrie, D.; Pudjiastuti, P. Wound Healing and Antioxidant Evaluations of Alginate from *Sargassum ilicifolium* and Mangosteen Rind Combination Extracts on Diabetic Mice Model. *Appl. Sci.* **2021**, *11*, 4651. https://doi.org/10.3390/app11104651

Academic Editor: Genoveffa Nuzzo

Received: 10 March 2021
Accepted: 11 May 2021
Published: 19 May 2021

Publisher's Note: MDPI stays neutral with regard to jurisdictional claims in published maps and institutional affiliations.

Abstract: A diabetic foot ulcer is an open wound that can become sore and frequently occurs in diabetic patients. Alginate has the ability to form a hydrophilic gel when in contact with a wound surface in diabetic patients. Xanthones are the main compounds of mangosteen rind and have antibacterial and anti-inflammatory properties. The purpose of this research was to evaluate the wound healing and antioxidants assay with a combination of alginate from *S. ilicifolium* and mangosteen rind combination extracts on a diabetic mice model. The characterization of alginate was carried out by size exclusion chromatography with multiple angle laser light scattering (SEC-MALLS) and thermogravimetric analysis (TGA). The M/G ratio of alginate was calculated by using proton nuclear magnetic resonance (^{1}H NMR). The antioxidant activity of mangosteen rind and the combination extracts was determined using the DPPH method. The observed parameters were wound width, number of neutrophils, macrophages, fibrocytes, fibroblasts, and collagen densities. The 36 male mice were divided into 12 groups including non-diabetic control (NC), diabetes alginate (DA), alginate–mangosteen (DAM), and diabetes control (DC) groups in three different groups by a histopathology test on skin tissue. The treatment was carried out for 14 days and mice were evaluated on Days 3, 7, and 14. The SEC-MALLS results showed that the molecular weight and dispersity index (Đ) of alginate were 2.77×10^4 Dalton and 1.73, respectively. The M/G ratio of alginate was 0.77 and described as single-stage decomposition based on TGA. Alginate, mangosteen rind extract, and their combination were divided into weak, medium, and strong antioxidant, respectively. The treatment of the DA and DAM groups showed a decrease in wound width and an increase in the number of fibrocytes, fibroblasts, and macrophages. The number of neutrophils decreased while the percentage of collagen densities increased for all the considered groups.

Keywords: alginate; *Sargassum ilicifolium*; mangosteen rind; wound healing; diabetic mice

1. Introduction

Diabetes mellitus (DM) is a metabolic disease caused by the disruption of the glucose metabolism in the body [1]. DM can cause changes in skin homeostasis resulting in changes in metabolism and several complications, such as vasculopathy and neuropathy [2]. Diabetes is usually accompanied by the emergence of foot ulcer disease or gangrene, palpable femoral and popliteal pulses, and the absence of foot pulses [3]. Foot ulcer disease and gangrene are cases of disease that often occur in people with DM and usually begin with the appearance of wounds. In general, there are many ways to treat open wounds in diabetic patients: one of them being the use of ointments as topical medication. Agrawal et al. (2014) reported that wounds can be healed using antibiotics, either used topically, directly on the wound, or orally [4]. Diabetic wound healing can be disrupted if the patient has poor blood sugar control and therapy, as well as the presence of bacteria on the wound surface that causes infection [5]. In fact, the problems that occur during the wound-healing process make the wound worsen, e.g., by increasing the reactive oxygen species (ROS) level and oxidative stress. Both of these problems can disrupt the wound-healing process, so the healing takes longer. In normal conditions, ROS such as hydrogen peroxide (H_2O_2) and superoxide act as cellular messengers to stimulate wound healing [6]. Increasing the level of ROS has a beneficial effect but, in some cases, can also cause tissue damage [7]. When diabetic complications occur, the ROS level and oxidative stress start causing cell death and tissue damage via several mechanisms [8]. Disproportion between antioxidants and ROS disturbed by a depletion of antioxidants or an accumulation of ROS causes oxidative stress [9]. This situation might be resolved through the application of antioxidants. Lobo et al. (2010) reported that antioxidant molecules contribute an electron to make a free radical become neutral, thus reducing its capacity for cellular damage through their free radical scavenging property [10].

Alginate is a natural polymer that contains β-D-mannuronic acid (M) and α-L-guluronic acid (G) blocks [11,12]. The main source of alginate is the various genera of brown seaweed [12]. Alginate as a topical medicinal material has been chosen because it is considered capable in maintaining the humidity around the wound, minimizing bacterial infections, and easing the wound-healing process [13]. Alginate has been used in several types of wounds, including pressure, diabetic, venous ulcers, and some open wounds due to its ability as a good absorbent and gel-forming agent [14,15].

Mangosteen rind has been a traditional medicine to treat skin trauma and infections for many decades [16]. Several studies have reported that xanthones (such as α-mangostin) are the main compounds found in mangosteen rind. Xanthone has antibacterial, anti-inflammatory, antioxidant, anticancer, and cardioprotective activities [17]. In the present paper, the histopathology of wound healing on a diabetic mice model by using a composite of alginate from *S. duplicatum* and okra combination extracts is described. Here, we report the wound-healing and antioxidant evaluations of alginate from *S. ilicifolium* and mangosteen rind combination extracts on a diabetic mice model.

2. Materials and Methods

2.1. Materials

S. ilicifolium samples were taken from Kei Island, Maluku, Indonesia in July 2018. Identification was performed at the Oceanographic Research Center, LIPI, Jakarta, Indonesia (Approval Reference Number: 1/3/18-id/2018). Mangosteen (*Garcinia mangostana*) was purchased from the Rungkut district (Surabaya, Indonesia). The species identification was carried out at the Biosystem Laboratory of the Department of Biology of Airlangga University.

2.2. Sodium Alginate Extraction

The *S. ilicifolium* was dried, cut into pieces, and grounded to powder. The dried *S. ilicifolium* was added in a Becker in 0.1% KOH, soaked for an hour, and filtered. Then, 1% HCl (1:30 *w/v*) was added to the sample and stirred for an hour, then it was filtered and washed using water until a neutral pH was reached. The sample was then soaked with 2% Na_2CO_3 at ±70 °C for two hours and filtered. The filtrate was added with 10% HCl to pH 2.8–3.2 or until a gel was formed and 2% Na_2CO_3 until neutral pH was added to the gel. Then, 4% NaOCl (1:2 *v/v*) was added to the mixture. The alginate extract was pipetted into 2-propanol solution in a 1:2 *v/v* ratio and slowly stirred until a sodium alginate fiber was formed. The sodium alginate fiber was dried and grounded to form sodium alginate powder.

2.3. Analysis of Sodium Alginate

2.3.1. Proton Nuclear Magnetic Resonance (^{1}H NMR) Spectroscopy

^{1}H NMR spectroscopy analyses were conducted on a JEOL JNM-ECS400A spectrometer (JEOL, Peabody, MA, USA) at a frequency of 400 MHz. D_2O was used as the NMR solvent if not otherwise detailed. All samples were freeze dried before analysis.

2.3.2. Size Exclusion Chromatography with Multi-Angle Laser Light Scattering (SEC-MALLS) Analysis

The molecular weight of alginate and its classification were analyzed using the gel permeation chromatography technique by using a Shimadzu HPLC system (Shimadzu UK Limited, Milton Keynes, UK) comprising a CBM-20A Controller, LC-20AD Pump with degasser, SIL-20A Autosampler, and SPD-20A detector; HELEOS-II light-scattering and Optilab rEx refractive index detectors were provided by Wyatt. A PL aquagel-OH mixed column (7.5 × 300 mm, 8 μm particle size; Agilent Technologies, Santa Clara, CA, USA) and a PL aquagel-OH mixed guard column (7.5 × 50 mm, 8 μm particle size; Agilent Technologies, Santa Clara, CA, USA) were used as the stationary phase. The mobile phase was a 50 mM sodium nitrate solution with a flow rate of 0.5 mL/min [11]. The alginate solution (1.5 mg mL^{-1}), dissolved in deionized water, was filtered through a nylon membrane (Whatman, UK) before analysis. The value of 0.165 was used as the dn dc^{-1} grade [18].

2.3.3. Thermogravimetric Analysis (TGA)

TGA was carried out on a PL Thermal Sciences STA 625 thermal analyzer (PL Thermal Science Limited, Surrey, UK). A total of 10 mg of accurately weighed sample in an aluminum sample cup was placed into the furnace with a N_2 flow of 100 mL min^{-1} and heated from room temperature to 625 °C at a heating rate of 10 °C min^{-1}. From the TGA profiles, the temperatures at 10% and 50% mass loss (TD_{10} and TD_{50}, respectively) were analyzed afterwards.

2.4. Extraction of Mangosteen Rind

The dry mangosteen rinds were mashed into small fragments. The fragments were then macerated using 96% ethanol for three days. Samples were filtered and concentrated under vacuum. The extract was freeze dried and kept in 4 °C for the next step.

2.5. Antioxidant Assay and Analysis of Total Phenolic Content

Antioxidant assays were performed on alginate from *S. ilicifolium*, mangosteen rind extract, and a combination of alginate–mangosteen rind extracts by the 2,2-diphenyl-1-picryl-hydrazyl-hydrate (DPPH) free radical method. The concentration of DPPH and stock solutions were 50.0 and 1000.0 μg/mL, respectively. The extracts were diluted in concentration variations of 200.0, 150.0, 125.0, 100.0, 75.0, 50.0, 35.0, 25.0, 15.0, 10.0, and 6.0 μg/mL of sample solution in methanol (200 μL) and 50 μL of DPPH was added into 96-well plates. Subsequently, the mixtures were incubated for 30 min in a dark chamber and the absorbance was measured at 517 nm on an ELISA plate reader.

The total phenolic content of alginate and mangosteen rind extract was determined by the colorimetric method, referring to the procedure of Malik et al. (2015) with some modifications and with gallic acid as standard [19]. The standard solutions and samples with concentrations of 10, 20, 30, 40, and 50 ppm were added with 0.4 mL of Folin–Ciocalteau reagent. After 4–8 min, 4.0 mL of 7% Na_2CO_3 solution was added and aquabidestilata was added to 10 mL, and then it was allowed to stand for 2 h at room temperature. The absorbance of solution was analyzed at a maximum wavelength of 744.8 nm; a calibration curve was made for the relationship between the concentration of gallic acid (μg/mL) and the absorbance.

2.6. Induction of Diabetic Mice

The BALB/c strains of adult male mice (*Mus musculus*) were obtained from the Faculty of Pharmacy, Airlangga University, Surabaya. The mice were 3–4 months old with a weight range of 20–35 g. Ethical clearance for treating animals was obtained from the Faculty of Veterinary Medicine, Airlangga University, Surabaya, Indonesia with License Reference Number: 2. KE. 049.04.2019.

The mice were treated in a 12 h light and 12 h dark lighting system for a couple of weeks. During acclimatization for the experiment, the mice were fed (oral dose of 0.3 mL) with lard three times per week during 3 weeks in order to achieve a high fat diet. Streptozotocin (STZ) was injected for 8 days by using the 30 mg/kg body weight using multiple low dose method of 0.15 mL intraperitoneally (i.p) to induce type II diabetes mellitus [20]. The mice were weighed before and after treatment. The level of blood glucose was measured after STZ induction at 3, 7, and 14 days of treatment.

2.7. Animal Grouping and Treatment

The group of mice (12 groups) were separated: (a) three groups for no treatment as normal (N) and (b) nine diabetic groups for treatment (D), consisting of three mice in each group for 3, 7, and 14 days of treatment [20]. The nine treatment groups were classified into three categories: (1) three groups for diabetic control (DC3, DC7, and DC14); (2) three groups for alginate treatment (DA3, DA7, and DA14); and (3) three groups for alginate–mangosteen rind extract treatment (DAM3, DAM7, and DAM14). A centimeter wound on the mice's *glutea* (buttocks) was generated. Each group was smeared with Vaseline (untreated and DC), vaseline–alginate (DA), and vaseline–alginate–mangosteen rind extract ointments (DAM). A single dose of 50 mg/kg body weight ointment was generated for treatment. The animal number replication was referred to the Federer formula (1967) [21].

2.8. Histopathology Analysis of Wound Healing

The histopathology experiments of wound healing were conducted at the Pathology Laboratory of the Faculty of Veterinary Medicine, Airlangga University. A microscope at 400 times magnification was used for the observation of fibroblasts, fibrocytes, macrophages, neutrophils, and collagen density. The width of the wound was analyzed using 40 times magnification under an optical microscope.

2.9. Statistical Analysis

The wound-healing measurements are illustrated as a mean ± standard error mean (SEM). The wound-healing data analysis was conducted by a normality and homogeneity test, one-way ANOVA, and Duncan test. If the results of normality and homogeneity were not qualified ($\alpha = 0.05$), non-parametric tests such as the Kruskal–Wallis and Mann–Whitney tests were applied. All the statistical analyses were calculated using an IBM with SPSS 20.0 software.

3. Results

3.1. Analysis of Sodium Alginate

3.1.1. Proton Nuclear Magnetic Resonance (^{1}H NMR) Analysis

The mannuronate-to-guluronate (M/G) ratio of alginate was determined by ^{1}H NMR. The spectrum of alginate is shown in Figure 1. The proton signals of alginate at a chemical shift in the 4.8–4.0 ppm range correspond to the G_2/M_2, while the signals at 5.8–5.0 ppm correspond to the G_1/M_1, respectively. The G_1 and M_1 are the most de-shielded protons (δ 5.8–5.1 and 5.5–5.1, respectively). M_1 is more shielded than G_1, because G_1 is located at an equatorial position in comparison to an axial one in M_1. The M/G ratio of alginate was 0.77 and was determined by the integration as previously reported [22].

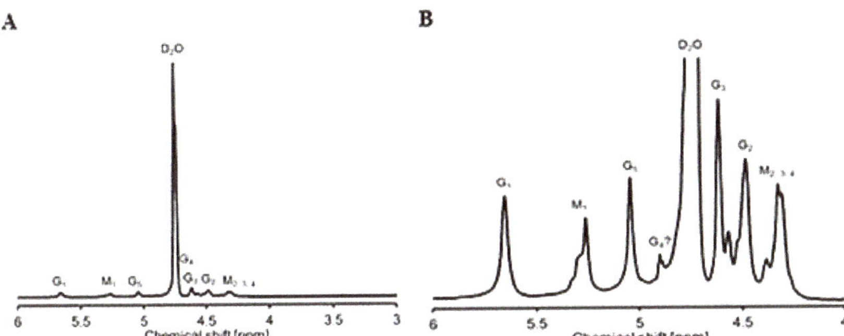

Figure 1. (**A**) Spectra of ^1H NMR spectra of alginate from *S. Ilicifolium* (90 °C 128 scans); (**B**) detail of ^1H NMR spectra (90 °C 128 sc).

3.1.2. Size Exclusion Chromatography with Multi-Angle Laser Light Scattering (SEC-MALLS) Analysis

The polydispersity index (Đ) and the molecular weight of alginate from *S. ilicifolium* were analyzed by SEC-MALLS (Figure 2A). The number average molecular weight (M_N) and weight average molecular weight (M_W) of alginate from *S. ilicifolium* were 1.49×10^4 and 2.77×10^4 Dalton, respectively. The Đ was found to be 1.73, which indicated that alginate from *S. ilicifolium* has a good homogeneity and is classified as chain-growth polymerization.

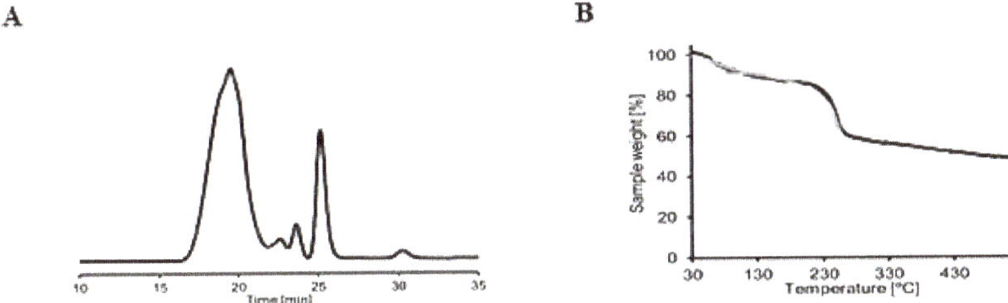

Figure 2. (**A**) RI signal of size exclusion chromatography with multiple angle laser light scattering (SEC-MALLS) analysis and (**B**) thermogravimetric analysis (TGA) of sodium alginate.

3.1.3. Thermogravimetric Analysis

Thermal stability and polymer matrix strength were analyzed using TGA by observing the weight difference as the sample was heated at a constant rate. Figure 2B shows the thermal stability and strength of the *S. ilicifolium* alginate polymer matrix. It was found that 5% of the mass was obscured (TD_5) at 82 °C and most probably indicated the water loss (moisture). The 10% (TD_{10}) and 20% mass loss occurred at 125 °C and 235 °C, respectively, indicating glycosidic bond destruction. The 50% (TD_{50}) mass loss at 466 °C indicated the carbon turns into charcoal.

3.2. Antioxidant Assay and Analysis of Total Phenolic Content

An antioxidant assay was utilized to measure the inhibitory concentration (IC_{50}) of each compound. The antioxidant activity of alginate, mangosteen rind, and the combination of alginate–mangosteen rind extracts was found using the DPPH method. The IC_{50} of alginate, mangosteen rind, and the combination of alginate–mangosteen rind extract was 297.60, 29.60, and 52.72 μg/mL, respectively (Table 1). The results showed that alginate was a moderate antioxidant, mangosteen rind was strong antioxidant, and the combination of alginate–mangosteen rind extracts was acting as a very powerful antioxidant. The total phenolic content in the mangosteen rind extract was 32.94 mg GAE/g extract or 3.29% (y = 0.0193x + 0.031; R^2 = 0.9976).

Table 1. Antioxidant activity of alginate, mangosteen rind, and alginate–mangosteen rind combination.

Sample	IC_{50} (μg/mL)	Classification
S. ilicifolium alginate	120.88	Moderate antioxidant
Mangosteen rind extract	29.06	Powerful antioxidant
Alginate–mangosteen rind extract	52.72	Strong antioxidant

3.3. Measurement of Body Weight in Mice

The administration of lard for three weeks in a single oral dose (0.3 mL) can significantly increase the body weight of mice ($p < 0.05$). The increase in the body weight of mice ranged from 29 ± 3.42 g to 34 ± 2.83 g. Husen et al. (2019) reported that obesity caused by excessive fat accumulation can induce various chronic diseases and complications such as diabetes mellitus [23].

3.4. Measurement of Blood Glucose Level in Mice

STZ can increase blood sugar levels of up to more than 250 mg/dL in mice, as presented in Table 2. Based on experiments, the blood sugar levels of the normal group decreased significantly on Days 1 to 14 ($p < 0.05$). The blood glucose level in the DC group decreased from Days 1 to 7 but increased on Day 14 (250.0 ± 10.6 mg/dL). The DA group showed a decrease from Days 1 to 14 (250.0 ± 8.5 mg/dL to 209.0 ± 11.3 mg/dL). Fluctuating changes in body weight were observed in the DAM mice group, but on Day 14 the blood sugar level decreased to 135.0 ± 7.1 mg/dL. On Day 14, the DAM mice showed a normal blood sugar level (non-diabetic).

Table 2. Mean and standard deviation of blood glucose.

Group Treatment	Blood Glucose Level (mg/dL)			
	Day 1	Day 3	Day 7	Day 14
Normal (NC)	118.0 ± 4.2	117.0 ± 2.3	114.0 ± 2.8	110.0 ± 1.4
Diabetic (DC)	256.0 ± 4.9	250.0 ± 11.3	242.0 ± 2.1	250.0 ± 10.6
Alginate ointment (DA)	250.0 ± 8.5	230.0 ± 12.0	225.0 ± 2.8	209.0 ± 11.3
Alginate–mangosteen rind extract ointment (DAM)	251.0 ± 5.7	176.0 ± 5.7	182.0 ± 7.8	135.0 ± 7.1 **

Asterisk notation in the table shows the result of the statistical test ($\alpha < 0.05$). ** notation shows insignificant differences with NC but significant differences with DC. ** = $0.01 \leq \alpha < 0.05$. $n = 3$ animals per time point.

3.5. Microscopic Evaluation of Wound Tissue

The microscopic evaluation of wound tissue included an evaluation of wound width, the number of neutrophil cells, macrophages, fibrocytes, and fibroblasts, and the collagen density in each group. The results of the histopathological microscopic observations in all study groups are described in Figures 3 and 4.

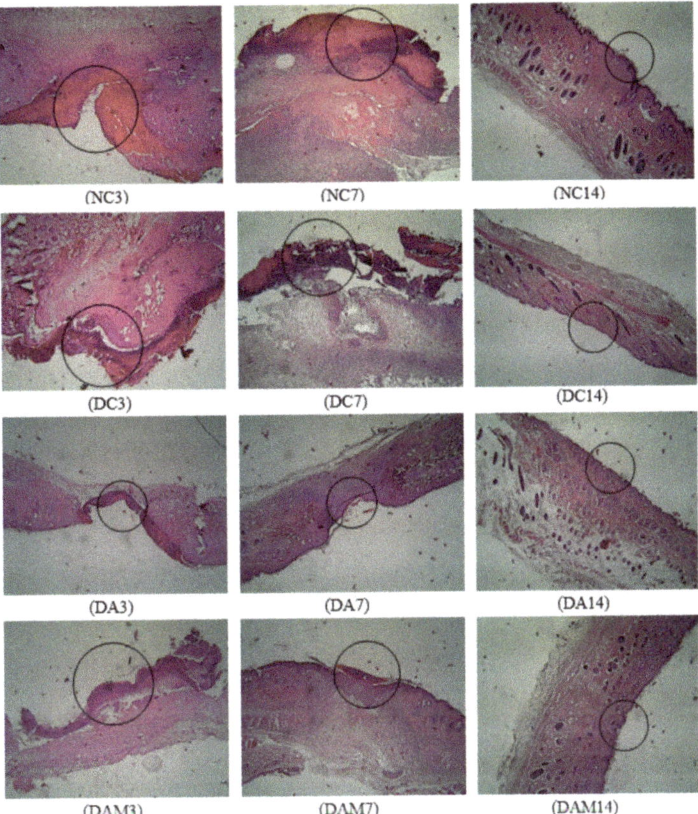

Figure 3. Histopathology analysis of wound width was measured on the three variation times. A circle in the picture indicates the wound width in the specimen. NC: normal group; DA: alginate treatment group; DAM: alginate–mangosteen rind extract treatment group; and DC: diabetic group.

Based on microscopic observations, it can be seen that the DAM group revealed the best wound-healing process for diabetic mice compared to other diabetic treatment groups. The re-epithelization of DA and DAM occurred on Day 7 and wound closing occurred on Day 14 for all groups. Figure 4 shows that the DAM group had the best re-epithelization process among the groups.

Figure 4 shows the wound parameters in microscopic observation, where neutrophils (green), macrophages (blue), fibrocytes (yellow), and fibroblasts (red) are shown by an arrow in a different color. The number of neutrophils increased at Day 3 and decreased at Day 7 and Day 14 for all groups. The number of macrophages increased at Day 3 to Day 7, and then declined at Day 14 for all groups. NC, DA, and DAM showed a significant rise in the number of fibrocytes and fibroblasts at Day 7 to indicate re-epithelization and a reduction at Day 14.

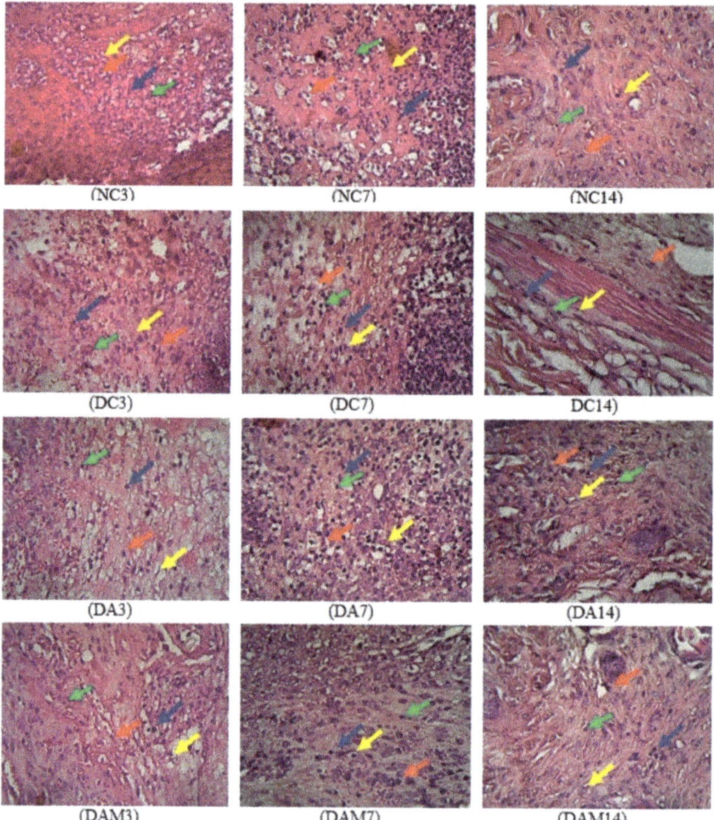

Figure 4. Analysis of wound width on the three variation times. Green arrows: neutrophils; blue arrows: macrophages; yellow arrows: fibrocytes; and red arrows: fibroblasts. NC: normal group; DA: alginate treatment group; DAM: alginate–mangosteen rind extract treatment group; and DC: diabetic group.

3.6. *Parameter Analysis*

3.6.1. Wound Width Determination

The width of the wound is defined as the distance between the epithelium (right–left) that underwent a complete re-epithelialization quantifiable in microns. The width of the wound for each group per day showed a significant difference (Figure 3). All groups showed good results on Day 14 as the wound had closed in all groups. DAM showed the best wound repair response in comparison with the NC group (2502.0 ± 10.4 μm and 726.0 ± 29.7 μm on Days 3 and 7, respectively) (Figure 5).

3.6.2. Neutrophil Measurement

Neutrophils are the onset of an inflammatory response in the recovery process of open wounds and aim to clean microorganisms that enter the wound [24–26]. The number of neutrophils showed significant differences both in the DA and DAM groups toward DC. A decrease in the neutrophil count was observed in DAM from Days 3 to 14, similar to what was observed for the NC group (Figure 6).

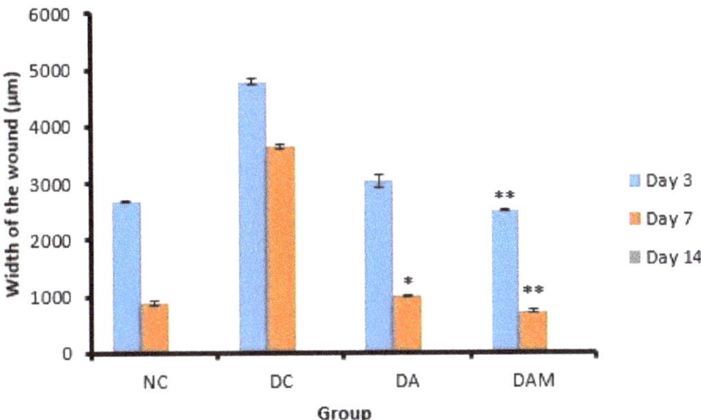

Figure 5. Wound width at the three variation times. Notation on the graph shows the result of the statistical test ($\alpha < 0.05$). * and ** notation shows insignificant differences with NC but significant differences with DC. * = $0.01 \leq \alpha < 0.05$; ** = $\alpha \leq 0.01$. NC: normal group; DA: alginate treatment group; DAM: alginate–mangosteen rind extract treatment group; and DC: diabetic group.

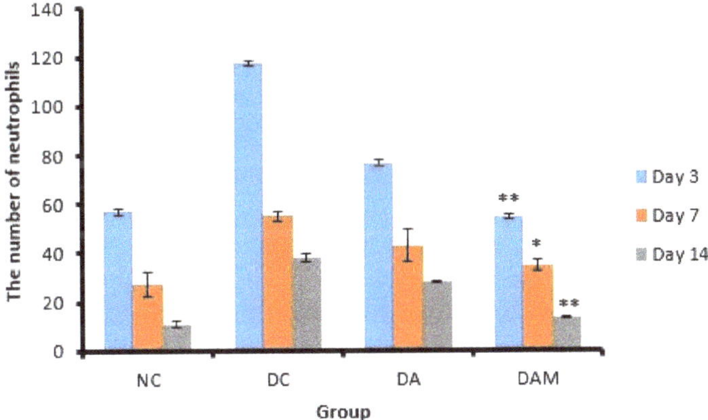

Figure 6. Number of neutrophils at the three variation times. Notation on the graph shows the result of the statistical test ($\alpha < 0.05$). * and ** notation shows insignificant differences with NC but significant differences with DC. * = $0.01 \leq \alpha < 0.05$; ** = $\alpha \leq 0.01$. NC: normal group; DA: alginate treatment group; DAM: alginate–mangosteen rind extract treatment group; and DC: diabetic group.

3.6.3. Macrophage Measurement

Macrophage secretion will increase massively as neutrophils decrease [26]. Figure 7 shows that the number of macrophages in the DC group showed a high increase due to hyperglycemia and ROS, while the other groups experienced a decrease because they did not experience these conditions. The DAM group gave a similar observation as the NC group, wherein the graph of macrophages in this group was the same as the control normal/non-diabetic group (Figure 7).

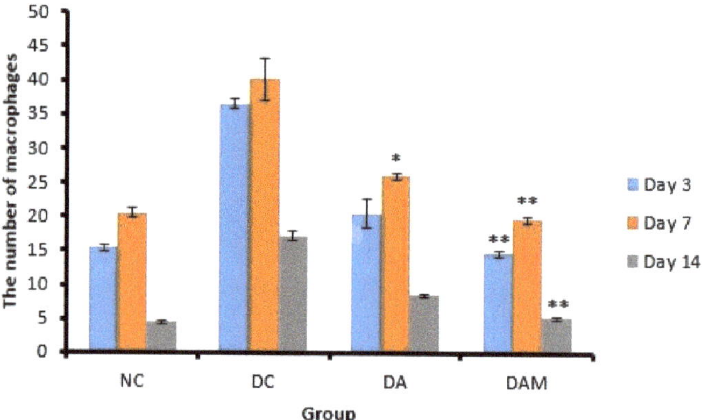

Figure 7. Number of macrophages at the three variation times. Notation on the graph shows the result of the statistical test ($\alpha < 0.05$). * and ** notation shows insignificant differences with NC but significant differences with DC. * = $0.01 \leq \alpha < 0.05$; ** = $\alpha \leq 0.01$. NC: normal group; DA: alginate treatment group; DAM: alginate–mangosteen rind extract treatment group; and DC: diabetic group.

3.6.4. Fibrocyte Measurement

Fibrocytes play a role in the process of differentiation of myofibroblasts in fibrosis and wound healing [27]. In this study, the number of fibrocytes in the DAM groups on Days 3, 7, and 14 were close to the NC group, whereas the DC group showed the lowest number of fibrocytes, i.e., almost one-third of the numbers recorded for the NC and DAM groups (Figure 8).

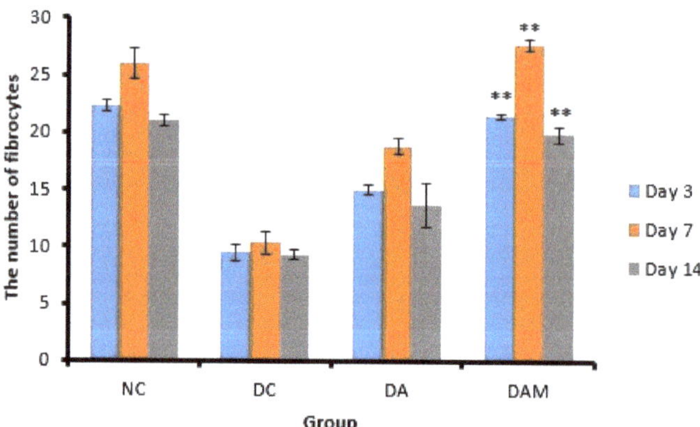

Figure 8. Number of fibrocytes at the three variation times. Notation on the graph shows the result of the statistical test ($\alpha < 0.05$).** notation shows insignificant differences with NC but significant differences with DC. ** = $\alpha \leq 0.01$. NC: normal group; DA: alginate treatment group; DAM: alginate–mangosteen rind extract treatment group; and DC: diabetic group.

3.6.5. Fibroblast Measurement

The histopathological observations are displayed in Figure 4. DAM showed the highest number of fibroblasts, while DC showed the lowest number of fibroblasts (Figure 9). The number of fibroblasts on Days 3 and 7 showed an increase but decreased on Day 14 in all groups (Figure 9).

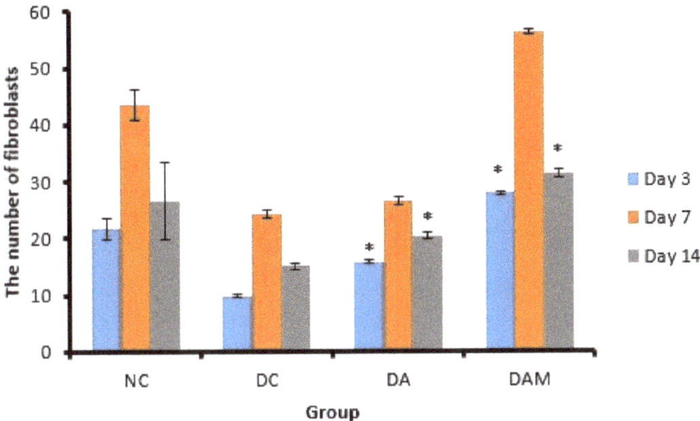

Figure 9. Number of fibroblasts at the three variation times. Notation on the graph shows the result of the statistical test ($\alpha < 0.05$). * notation shows insignificant differences with NC but significant differences with DC. * = $0.01 \leq \alpha < 0.05$. NC: normal group; DA: alginate treatment group; DAM: alginate–mangosteen rind extract treatment group; and DC: diabetic group.

3.6.6. Collagen Density Measurement

Observation of collagen density was conducted to determine the process of the formation of new collagen in open wounds. Figure 10 shows that DAM presents the highest collagen density compared to the other groups, while DC shows the lowest percentage of collagen density among the considered groups.

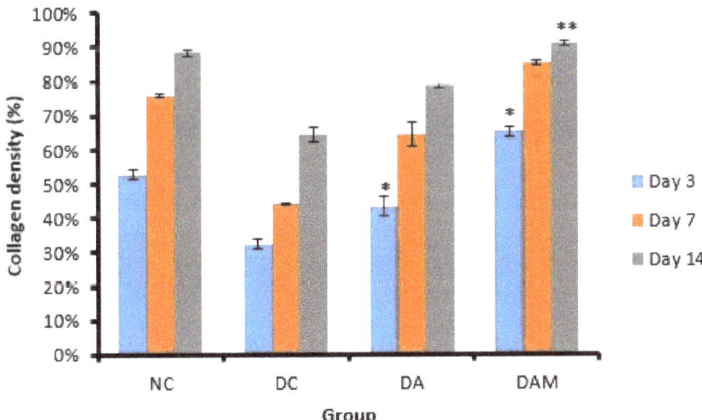

Figure 10. Collagen density was evaluated at the three variation times. Notation on the graph shows the result of the statistical test ($\alpha < 0.05$). * and ** notation shows insignificant differences with NC but significant differences with DC. * = $0.01 \leq \alpha < 0.05$; ** = $\alpha \leq 0.01$. NC: normal group; DA: alginate treatment group; DAM: alginate mangosteen rind extract treatment group; and DC: diabetic group.

4. Discussion

Sodium alginate was extracted from *S. ilicifolium* to give a pale-yellow solid with a 14.36% yield. The obtained yield is very similar to the one reported by Latifi et al. (2015), who reported yields ranging from 12% to 16.5% [28]. Davis et al. (2004) extracted *S. fluitans* and *S. oligocystum* using high temperature and base conditions to obtain 21.1–22.8% and 18.9–20.5% yields, respectively [29]. Mushollaeni (2011) reported that alginate from several species of *Sargassum* sp. had 16.93 to 30.50% yields [30]. Based on this study, the yield of the extracted alginate not only depends on the extraction method and the *Sargassum* species used but also on the seasonal variation. It is in fact well known that there are marked seasonal variations of polysaccharide contents in seaweeds, as polysaccharides are stored in winter as a food source and therefore a harvest at the end of summer will give higher polysaccharide levels than in the spring.

The ^{1}H NMR analysis of *S. ilicifolium* alginate showed that the signals in the range 4–4.8 ppm belong to protons from the β-glycosidic bond, whereas the signals between 5.1 and 5.8 ppm are resonances of protons from the α-glycosidic bond. The identification was performed by comparison with the ^{1}H NMR spectrum of alginate from Llanes et al. (1997) [31]. The structure of alginate from *S. ilicifolium* consists of β-D-mannuronic acid (4.3–4.8 ppm) and α-L-guluronic acid (5.1–5.8 ppm). The M/G ratio of alginate was determined by the integration of each proton with their monomers. The M/G ratio of alginate was 0.77, indicating that *S. ilicifolium* alginate has a higher guluronic acid monomer than mannuronic acid. The protons G_1 and M_1 are the most de-shielded protons (δ 5.1–5.8 and 5.1–5.5, respectively). M_1 is more shielded than G_1, because G_1 is located at an equatorial position in comparison to an axial one in M_1. In fact, Llanes et al. performed a pretreatment to partly hydrolyze the alginate before ^{1}H NMR analysis, while in our case we decided to work with non-hydrolyzed alginate as it was fully soluble in the NMR solvent. This can be an explanation for the slightly more structured peaks and possibly also for the small shifts downfield we observed.

The number of M_n and M_w of alginate were 1.49×10^4 and 2.77×10^4 Dalton, respectively. Furthermore, the M_n and M_w obtained were used in the calculation of the dispersity index (Đ) using the following equation:

$$Đ = Mw/Mn \quad (1)$$

The dispersity index (Đ) was 1.73. The Đ value of alginate from *S. ilicifolium* was below 2, indicating that the extracted alginate showed good homogeneity and was classified as chain-growth polymerization. Chain-growth polymerization is a process in which the high molecular weight polymer is generated at the beginning of the polymerization process and the polymer yield (the percentage conversion of the monomer into the polymer) gradually increases with time [32]. Based on TGA, alginate from *S. ilicifolium* had a single-stage decomposition curve (Figure 2B).

The antioxidant assay showed that the IC_{50} of alginate, mangosteen rind, and alginate–mangosteen rind extract combination was 297.60, 29.60, and 52.0 µg/mL, respectively. The results indicated that alginate was a weak antioxidant and mangosteen rind was a powerful antioxidant, and when combined they produced a material with excellent antioxidant properties. These observations could be linked to the fact that mangosteen rind contains xanthones, such as α-mangostin and γ-mangostin, which have been reported to exhibit anti-inflammatory, antioxidant, and radical scavenging activity [33–35]. According to Thong et al. (2015), xanthones can capture free radicals in two ways: through hydrogen atom transfer (HAT) to explain the antioxidant activity in the gas phase and the single electron transfer–proton transfer (SETPT) mechanism that is thermodynamically favored in water [36]. Strong antioxidant activity observed in the combination of alginate and mangosteen rind extract could possibly overcome ROS that occur in hyperglycemic conditions, thus enhancing wound repair in diabetic mice.

The total phenolic content in mangosteen rind extract tends to be low due to several factors, such as the type of mangosteen fruit, storage time, and the hardness of the rind. According to Dangcham et al. (2008), the total phenolic content of mangosteen fruit will decrease throughout the storage time [37]. Another study reported that the lignin content and density of the damaged pericarp tissue would increase as a result of the hardening of the fruit skin, while the total phenolic content decreased [38]. Mangosteen rind is an essential, natural phenolic antioxidant source; at least ten phenolic acids have been characterized in mangosteen rind [39,40]. Mangosteen rind contains bioactive substances, such as phenolic acids, flavonoids, anthocyanins, proanthocyanidins, (-)-epicatechin, and xanthones, which have biological, medical, and antioxidant properties.

The administration of lard can significantly increase the body weight of mice. This condition indicated increasing hyperglycemia and insulin resistance, and it will cause type II DM [20]. Husen et al. (2019) reported that obesity caused by excessive fat accumulation can induce various chronic diseases and complications such as diabetes mellitus [23]. The administration of STZ was performed to increase the condition of hyperglycemia (increased blood sugar level) [20]. The increase of blood sugar levels in three diabetic groups, namely diabetic non-treated (DC), diabetic treated with alginate ointment (DA), and diabetic treated in alginate–mangosteen rind extract ointment (DAM), of up to 250 mg/dL indicated that the mice were in a diabetic condition. Based on experiments, the combination of alginate and mangosteen rind extract proved able to reduce blood sugar levels (Table 2). Lee et al. (2018) reported that mangosteen was able to increase insulin secretion in pancreatic B-cells and protected cells from apoptosis, due to the presence of α-mangostin in mangosteen fruit [41]. Jariyapongskul et al. (2015) reported that daily α-mangostin supplementation in diabetic rats shows remarkable hypoglycemic and insulinotropic effects, a reduction in plasma glycated hemoglobin, and a decrease in serum triglycerides and cholesterol [42].

It was expected that the administration of a topical combination of alginate and mangosteen rind extract in diabetic wounds could reduce blood sugar levels by entering the systemic circulation through the bloodstream. The entry of compounds found in topical drugs through the bloodstream can enhance the host insulin level and the sensitivity of somatic cells [43]. Therefore, the topical combination of alginate and mangosteen rind extract can be classified as diadermic ointment. A diadermic ointment is a drug that is able to enter the deepest skin tissue and enter the systemic circulation [44]. However, further investigation is required to confirm whether the compounds responsible for reducing the blood sugar levels are a combination of alginate and mangosteen rind extract or the presence of other active compounds that have not been separated from the alginate during extraction given the number of weak proton signals observed in the ^{1}HNMR. In addition, it is necessary to confirm the exact mechanism concerning how blood glucose levels can be decreased by administering a topical combination of alginate and mangosteen rind extract.

The DAM group, which was treated by alginate–mangosteen rind extract combination, showed a very significant decrease on Days 3 and 7, 2502.0 $\pm$ 10.4 and 726.0 $\pm$ 29.7, respectively, compared to other groups (Table 3). On Day 14, all groups showed that the wound had closed. The wound width of diabetic mice in the DAM group showed the best activity when compared to the other tested conditions. This happened due to the combination of alginate and mangosteen rind extract providing strong antioxidant activity and thus decreasing ROS and increasing wound contraction and wound healing.

The diabetic group (DC) showed the greatest increase in the number of neutrophils in comparison to other groups. Administration of the combined alginate and mangosteen rind extracts (DAM) demonstrated that the neutrophil count in open wounds on diabetic mice increased in the first 24–48 h and then decreased until Day 14 (55.0 $\pm$ 0.9 on Day 3 to 14.0 $\pm$ 0.2 cells/mm^2 on Day 14), thus indicating that DAM could ease the process of wound healing (Table 4). The decrease in the neutrophil count in the DAM group occurred due to the presence of phenolic compounds from the mangosteen rind extract that acted as antioxidants. Phenolic compounds are able to reduce the secretion of human neutrophil elastase (HNE), which is an enzyme that can break down the components of the extracellular matrix by reducing the number of neutrophils [45].

The macrophage count in the DC group showed a slight decrease on Day 14 compared to the other groups, thus implying that the inflammation phase in the open wounds of diabetic mice occurred for a longer duration than the normal group. Persistent inflammatory conditions inhibit wound healing [46]. The presence of hyperglycemia and oxidative stress (increased blood sugar levels and ROS) can affect the modulation and polarization of macrophages, which could inhibit the healing process [27,47]. Therefore, the administration of alginate and mangosteen rind extracts to diabetic wounds could reduce the number of macrophages in the wound almost as well as non-diabetic wounds (Table 5).

Table 3. Counts considering the mean and standard deviation of wound width.

Group Treatment	Wound Width (μm)		
	Day 3	Day 7	Day 14
Normal (NC)	2673.0 ± 22.9	882.0 ± 44.1	0.0 ± 0.0
Diabetic (DC)	4786.0 ± 63.9	3636 ± 37.0	0.0 ± 0.0
Alginate ointment (DA)	3030.0 ± 126.6	1002.0 ± 11.5 *	0.0 ± 0.0
Alginate–mangosteen rind ointment (DAM)	2502.0 ± 10.4 **	726.0 ± 29.7 **	0.0 ± 0.0

Asterisk notation in the table shows the result of the statistical test ($\alpha < 0.05$). * and ** notation shows insignificant differences with NC but significant differences with DC. * = $0.01 \leq \alpha < 0.05$; ** = $\alpha \leq 0.01$.

Table 4. Cell counts considering the mean and standard deviation of neutrophils.

Group Treatment	Neutrophil (cells/mm^2)		
	Day 3	Day 7	Day 14
Normal (NC)	57.0 ± 1.4	28.0 ± 4.9	11.0 ± 1.4
Diabetic (DC)	118.0 ± 1.2	55.0 ± 1.9	38.0 ± 1.4
Alginate ointment (DA)	77.0 ± 1.2	43.0 ± 6.4	28.0 ± 0.4
Alginate–mangosteen rind ointment (DAM)	55.0 ± 0.9 **	35.0 ± 2.1 *	14.0 ± 0.2 **

Asterisk notation in the table shows the result of the statistical test ($\alpha < 0.05$). * and ** notation shows insignificant differences with NC but significant differences with DC. * = $0.01 \leq \alpha < 0.05$; ** = $\alpha \leq 0.01$.

Table 5. Cell counts considering the mean and standard deviation of macrophages.

Group Treatment	Macrophage (cells/mm^2)		
	Day 3	Day 7	Day 14
Normal (NC)	15.0 ± 0.5	21.0 ± 0.7	5.0 ± 0.2
Diabetic (DC)	37.0 ± 0.7	40.0 ± 3.1	17.0 ± 0.7
Alginate ointment (DA)	21.0 ± 2.1	26.0 ± 0.5 *	9.0 ± 0.2
Alginate–mangosteen rind ointment (DAM)	15.0 ± 0.5 **	20.0 ± 0.5 **	5.0 ± 0.2 **

Asterisk notation in the table shows the result of the statistical test ($\alpha < 0.05$). * and ** notation shows insignificant differences with NC but significant differences with DC. * = $0.01 \leq \alpha < 0.05$; ** = $\alpha \leq 0.01$.

An increase in the fibrocyte count leads to an increase in collagen V production, while lowering collagen I, III, and IV levels [48]. Fibrocyte differentiation can reduce inflammatory conditions and tissue damage as well as improve the wound-healing and tissue-remodeling process. The number of fibrocytes will increase with the increasing age of the wound, and more than 15 fibrocytes indicated the age of the wound was between 9 and 14 days [27]. The administration of a topical combination of alginate and mangosteen rind extract in the DAM group can increase the number of fibrocyte cells in diabetic open wounds between Day 3 and Day 7, thus accelerating the process of angiogenesis. On the other hand, the fibrocyte count decreased by Day 14, as shown in Table 6; this condition indicated the wound had healed.

Table 6. Cell counts considering the mean and standard deviation of fibrocytes.

Group Treatment	Fibrocyte (cells/mm^2)		
	Day 3	Day 7	Day 14
Normal (NC)	22.0 ± 0.5	26.0 ± 0.4	21.0 ± 0.5
Diabetic (DC)	10.0 ± 0.7	10.0 ± 0.9	9.0 ± 0.5
Alginate ointment (DA)	15.0 ± 0.5	19 ± 0.7	14.0 ± 1.9
Alginate–mangosteen rind ointment (DAM)	22.0 ± 0.2 **	28 ± 0.5 **	20.0 ± 0.7 **

Asterisk notation in the table shows the result of the statistical test ($\alpha < 0.05$). ** notation shows insignificant differences with NC, but significant differences with DC. ** = $\alpha \leq 0.01$.

Fibroblasts are cells that have a role in the proliferation phase. These cells have the function to regulate collagen, glycosaminoglycan, proteoglycan, fibronectin, and elastin as the extracellular matrix components [49,50]. In addition, fibroblasts also play a role in collagen. The presence of an increase and decrease in the number of fibroblasts is the same as observed in the measurement of the number of fibrocytes in the wound. All groups showed an increase in the number of fibroblasts on Day 7 and a decline on Day 14. A decrease in the number of fibroblasts on Day 14 in all groups indicates that the healing phase is at the remodeling stage (Table 7). The remodeling phase can be identified by decreasing proliferation and inflammation, reorganization of the extracellular matrix, and regression of newly formed capillary vessels [51]. The fibroblast count in the DAM group was almost the same as the fibroblast count of the NC or normal group. This result indicated that the administration of alginate–mangosteen rind extract combination was effective to heal the open wound.

Table 7. Cell counts considering the mean and standard deviation of fibroblasts.

Group Treatment	Fibroblast (cells/mm^2)		
	Day 3	Day 7	Day 14
Normal (NC)	22.0 ± 1.9	44.0 ± 2.6	27.0 ± 6.8
Diabetic (DC)	10.0 ± 0.2	24.0 ± 0.7	15.0 ± 0.5
Alginate ointment (DA)	16.0 ± 0.2 *	27.0 ± 0.7	20.0 ± 0.5 *
Alginate–mangosteen rind ointment (DAM)	28.0 ± 0.2 *	56.0 ± 0.5	31.0 ± 0.7 *

Asterik notation in the table shows the result of the statistical test ($\alpha < 0.05$). * notation shows insignificant differences with NC but significant differences with DC. * = $0.01 \leq \alpha < 0.05$.

The treatment group of alginate and mangosteen rind extract of DAM showed significant results at each observation on Days 3, 7, and 14 with 65.0 ± 1.4%, 85.0 ± 0.7%, and 91.0 ± 0.7%, respectively (Table 8). The DAM group showed the highest percentage of collagen density compared to other groups. The diabetic conditions increase the production of ROS/RNS, which decreases the collagen synthesis [52]. An increase in the percentage of collagen density in the DAM indicated an increase in collagen synthesis.

Table 8. Cell counts considering the mean and standard deviation of collagen density.

Group Treatment	Collagen Density (%)		
	Day 3	Day 7	Day 14
Normal (NC)	53.0 ± 1.4	76.0 ± 0.5	88.0 ± 0.9
Diabetic (DC)	32.0 ± 1.4	44.0 ± 0.2	65.0 ± 2.1
Alginate ointment (DA)	43.0 ± 2.8 *	65.0 ± 3.5	79.0 ± 0.7
Alginate–mangosteen rind ointment (DAM)	65.0 ± 1.4 *	85.0 ± 0.7	91.0 ± 0.7 **

Asterisk notation in the table shows the result of the statistical test ($\alpha < 0.05$). * and ** notation shows insignificant differences with NC but significant differences with DC. * = $0.01 \leq \alpha < 0.05$; ** = $\alpha \leq 0.01$.

Achievement of the wound-healing process that occurred in DAM could be due to compounds possessing antioxidant activity from the combined mangosteen peel. Moreover, Kataria et al. (2014) reported that alginate is a good absorbent and gel formation agent that has homeostatic properties [15]. Alginate has been used in several types of wounds, including pressure wounds, diabetic wounds, and venous ulcers such as cavities and multiple bleeding sores [14]. In addition, alginate acts as an antioxidant that can capture free radicals and absorbs wound exudate.

5. Conclusions

The administration of a topical combination of alginate extracted from *S. ilicifolium* and mangosteen rind extract (*Garcinia mangostana*) in diabetic open wounds is able to accelerate the process of wound closure. This is evidenced with an increase in neutrophils, macrophages, fibrocytes, fibroblasts, and collagen density, which enhance the process of re-epithelialization and the synthesis of collagen in the wound area. Alginate absorbs wound exudate, and xanthones are known to have antioxidant and anti-inflammatory activities. Xanthone is capable of capturing free radicals (ROS/RNS), which usually tend to increase in diabetics. Therefore, it can be concluded that the combination of alginate and mangosteen rind extract can improve the wound-healing process in diabetic mice.

Author Contributions: This paper is part of a master's degree thesis. Conceptualization, P.P. and D.W.; methodology, P.A.C.W. and S.A.H.; software, P.A.C.W.; validation, D.W., A.P., and S.A.H.; formal analysis, P.A.C.W., A.P., and M.V.; investigation, P.A.C.W.; resources, M.A.A.; data accuracy, P.A.C.W.; writing—original draft preparation, P.A.C.W.; writing—review and editing, P.P., K.A., and D.M.; visualization, P.A.C.W.; supervision, P.P. and D.W.; project administration, Z.N.I. and S.A.H.; funding acquisition, P.P. All authors have read and agreed to the published version of the manuscript.

Funding: This research was supported by the Innovation and Research Center, Airlangga University, Grant Number 1408/UN3/2019, and the APC was funded by the Innovation and Research Center, Airlangga University.

Institutional Review Board Statement: The study was conducted according to the guidelines of the Declaration of Helsinki, and approved by the Institutional Review Board (or Ethics Committee) of ANIMAL CARE AND USE COMMITTEE, Airlangga University (No.2.KE.049.04.2019, 4 April 2019).

Informed Consent Statement: Not applicable.

Data Availability Statement: The data presented in this study are available on request from the corresponding author, upon reasonable request.

Acknowledgments: The author would like to thank the Innovation and Research Center, Airlangga University for funding this research through the Mandat Research Grant, Airlangga University FY 2019 and Leonardo Gomez from the CNAP of the University of York for granting access to the SEC-MALLS system.

Conflicts of Interest: The author declares no conflict of interest.

References

1. Alghobashy, A.A.; Alkholy, U.M.; Talat, M.; Abdalmonem, N.; Zaki, A.; Ahmed, I.A.; Mohamed, R.H. Trace elements and oxidative stress in children with type 1 diabetes mellitus. *Diabetes Metab Syndr. Obes.* **2018**, *11*, 85–92. [CrossRef] [PubMed]
2. Behm, B.; Schreml, S.; Landthaler, M.; Babilas, P. Skin signs in diabetes mellitus. *J. Eur. Acad. Dermatol. Venereol.* **2012**, *26*, 1203–1211. [CrossRef] [PubMed]
3. Mills, J.L. Lower limb ischaemia in patients with diabetic foot ulcers and gangrene: Recognition, anatomic patterns and revascularization strategies. *Diabetes Metab Res. Rev.* **2016**, *32* (Suppl. S1), 239–245. [CrossRef]
4. Agrawal, P.; Soni, S.; Mittal, G.; Bhatnagar, S. Role of polymeric biomaterials as wound healing agents. *Int. J. Low. Extrem. Wounds* **2014**, *13*, 180–190. [CrossRef] [PubMed]
5. Bergin, S.; Wraight, P. Silver based wound dressings and topical agents for treating diabetic foot ulcers (Review). *Cochrane Libr.* **2006**. [CrossRef]
6. Guo, S.; DiPietro, L.A. Factors Affecting Wound Healing. *J. Dent. Res.* **2010**, *89*, 219–229. [CrossRef]
7. Rodriguez, P.G.; Felix, F.N.; Woodley, D.T.; Shim, E.K. The role of oxygen in wound healing: A review of the literature. *Derm. Surg.* **2008**, *34*, 1159–1169. [CrossRef]
8. Volpe, C.M.O.; Villar-Delfino, P.H.; dos Anjos, P.M.F.; Nogueira-Machado, J.A. Cellular death, reactive oxygen species (ROS) and diabetic complications. *Cell Death Dis.* **2018**, *9*, 119. [CrossRef] [PubMed]
9. Birben, E.; Sahiner, U.M.; Sackesen, C.; Erzurum, S.; Kalayci, O. Oxidative Stress and Antioxidant Defense. *World Allergy Organ. J.* **2012**, *5*, 9–19. [CrossRef]
10. Lobo, V.; Patil, A.; Phatak, A.; Chandra, N. Free radicals, antioxidants and functional foods: Impact on human health. *Pharm. Rev.* **2010**, *4*, 118–126. [CrossRef]
11. Pawar, S.N.; Edgar, K.J. Alginate Derivatization: A Review of Chemistry, Properties and Applications. *Biomaterials* **2012**, *33*, 3279–3305. [CrossRef] [PubMed]
12. Hay, L.D.; Rehman, Z.U.; Moradali, M.F.; Wang, Y.; Rehm, B.H.A. Microbial Alginate Production, Modification, and its Applications. *Microb. Biotechnol.* **2013**, *6*, 637–650. [CrossRef] [PubMed]
13. Lee, K.Y.; Mooney, D.J. Alginate: Properties and Biomedical Applications. *Prog. Polym. Sci.* **2012**, *37*, 106–126. [CrossRef] [PubMed]

14. Boateng, J.; Catanzano, O. Advanced Therapeutic Dressings for Effective Wound Healing—A Review. *J. Pharm Sci.* **2015**, *104*, 3653–3680. [CrossRef] [PubMed]
15. Kataria, K.; Gupta, A.; Rath, G.; Mathur, R.B.; Dhakate, S.R. In vivo wound healing performance of drug loaded electrospun composite nanobers transdermal patch. *Int. J. Pharm.* **2014**, *469*, 102–110. [CrossRef] [PubMed]
16. Kumar, S.; Bajwa, B.S.; Kuldeep, S.; Kalia, A.N. Anti-inflammatory activity of herbal plants: A review. *Int. J. Adv. Pharm. Biol. Chem.* **2013**, *2*, 272–281.
17. Fang, Y.; Su, T.; Qiu, X.; Mao, P.; Xu, Y.; Hu, Z.; Zhang, Y.; Zheng, X.; Xie, P.; Liu, Q. Protective effect of alpha-mangostin against oxidative stress induced-retinal cell death. *Sci. Rep.* **2016**, *6*, 210–218. [CrossRef] [PubMed]
18. Rinaudo, M. Main properties and current applications of some polysaccharides as biomaterials. *Polym. Int.* **2008**, *3*, 397–430. [CrossRef]
19. Malik, A.; Ahmad, A.R. Determination of phenolic and flavonoid contents of ethanolic extract of kanunang leaves (*Cordia myxa L.*). *Int. J. Pharm. Tech. Res.* **2015**, *7*, 243–246.
20. Novelli, M.; Bonamassa, B.; Masini, M.; Funel, N.; Canistro, D.; De Tata, V.; Martano, M.; Soleti, A.; Campani, D.; Paolini, M.; et al. Persistent correction of hyperglycemia in streptozotocin-nicotinamide-induced diabetic mice by a non-conventional radical scavenger. *Naunyn Schmiedebergs Arch. Pharm.* **2010**, *382*, 127–137. [CrossRef]
21. Federer, W.T. *Experimental Design, Theory and Application*; Oxford and IBH Publishing, Co.: New Delhi, India, 1967.
22. Ilmi, Z.N.; Wulandari, P.A.C.; Husen, S.A.; Winarni, D.; Alamsjah, M.A.; Awang, K.; Vastano, M.; Pellis, A.; Macquarrie, D.; Pudjiastuti, P. Characterization of Alginate from *Sargassum duplicatum* and the Antioxidant Effect of Alginate–Okra Fruit Extracts Combination for Wound Healing on Diabetic Mice. *Appl. Sci.* **2020**, *10*, 6082. [CrossRef]
23. Husen, S.A.; Wahyuningsih, S.P.A.; Ansori, A.N.M.; Hayaza, S.; Susilo, R.J.K.; Darmanto, W.; Winarni, D. The Effect of Okra (*Abelmoschus esculentus Moench*) Pods Extract on Malondialdehyde and Cholesterol Level in STZ-Induced Diabetic Mice. *Ecol. Environ. Conserv.* **2019**, *25*, 50–56.
24. Gillitzer, R.; Goebeler, M. Chemokines in cutaneous wound healing. *J. Leukoc. Biol.* **2001**, *69*, 513–521. [PubMed]
25. Artlett, C.M. Inflammasomes in wound healing and fibrosis. *J. Pathol.* **2013**, *229*, 157–167. [CrossRef]
26. Zhang, Y.; Li, L.; Liu, Y.; Liu, Z. PKM2 released by neutrophils at wound site facilitates early wound healing by promoting angiogenesis. *Wound Repair Regen.* **2016**, *24*, 328–336. [CrossRef]
27. Kondo, T.; Ishida, Y. Molecular pathology of wound healing. *Forensic Sci. Int.* **2010**, *203*, 93–98. [CrossRef] [PubMed]
28. Latifi, A.M.; Nejad, E.S.; Babavalian, H. Comparison of extraction of different methods of sodium alginate from brown algae *Sargassum sp.* localized in the southern of Iran. *J. Appl. Biotechnol. Rep.* **2015**, *2*, 251–255.
29. Davis, T.A.; Alfonso, M.R.; Larsen, M.B. Extraction, isolation and cadmium binding of alginate from *Sargassum* spp. *J. Appl. Phycol.* **2004**, *16*, 275–284. [CrossRef]
30. Mushollaeni, W. The physicochemical characteristics of sodium alginate from Indonesian brown seaweeds. *Afr. J. Food Sci.* **2011**, *5*, 142–149.
31. Llanes, F.; Sauriol, F.; Morin, F.O.; Perlin, A.S. An examination of sodium alginate from Sargassum by NMR spectroscopy. *Can. J. Chem.* **1997**, *75*, 585–590. [CrossRef]
32. Ramkumar, J. Nafion Perfluorosulphonate Membrane: Unique Properties and Various Applications. *Funct. Mater.* **2012**, 549–577.
33. Bumrungpert, A.; Kalpravidh, R.W.; Chuang, C.; Overman, A.; Martinez, K.; Kennedy, A.; McIntosh, M. Xanthones from Mangosteen Inhibit Inflammation in Human Macrophages and Human Adipocytes Exposed to Macrophage-Conditioned Media. *J. Nutr.* **2010**, *140*, 842–847. [CrossRef]
34. Sampath, P.D.; Vijayaraghavan, K. Cardioprotective effect of mangostin, a xanthone derivative from mangosteen on tissue defense system against isoproterenolinduced myocardial infarction in rats. *J. Biochem. Mol. Toxicol.* **2007**, *21*, 336–339. [CrossRef]
35. Chang, H.F.; Wu, C.H.; Yang, L. Antitumour and free radical scavenging effects of -mangostin isolated from *Garcinia mangostana* pericarps agains the patocellular carcinoma cell. *J. Pharm. Pharm.* **2013**, *65*, 1419–1428. [CrossRef] [PubMed]
36. Thong, N.M.; Quang, D.T.; Bui, N.H.T.; Dao, D.Q.; Nam, P.C. Antioxidant properties of xanthones extracted from the pericarp of *Garcinia mangostana* (Mangosteen): A theoretical study. *Chem. Phys. Lett.* **2015**, *625*, 30–35. [CrossRef]
37. Dangcham, S.; Bowen, J.; Ferguson, I.B.; Ketsa, S. Effect of temperature and low oxygen on pericarp hardening of mangosteen fruit stored at low temperature. *Postharvest Biol. Technol.* **2008**, *50*, 37–44. [CrossRef]
38. Bunsiri, A.; Ketsa, S.; Paull, R.E. Phenolic metabolisms and lignin synthesis in damaged pericarp of mangosteen fruit after impact. *Postharvest Biol. Technol.* **2003**, *29*, 61–67. [CrossRef]
39. Suttirak, W.; Manurakchinakorn, S. In vitro antioxidant properties of mangosteen peel extract. *J. Food Sci. Technol.* **2012**, *51*, 3546–3558. [CrossRef]
40. Zadernowski, R.; Czaplicki, S.; Naczk, M. Phenolic acid profiles of mangosteen fruits (*Garcinia mangostana*). *Food Chem.* **2009**, *112*, 685–689. [CrossRef]
41. Lee, D.; Kim, Y.M.; Jung, K.; Chin, Y.W.; Kang, K. Alpha-Mangostin Improves Insulin Secretion and Protects INS-1 Cells from Streptozotocin-Induced Damage. *Int. J. Mol. Sci.* **2018**, *19*, 1484. [CrossRef]
42. Jariyapongskul, A.; Areebambud, C.; Suksamrarn, S.; Mekseepralard, C. Alpha-mangostin attenuation of hyperglycemia-induced ocular hypoperfusion and blood retinal barrier leakage in the early stage of type 2 diabetes rats. *BioMed Res. Int.* **2015**, *2015*, 785826. [CrossRef] [PubMed]

43. Tan, W.S.; Arulselvan, P.; Ng, S.; Taib, C.N.M.; Sarian, M.N.; Fakurazi, S. Improvement of diabetic wound healing by topical application of Vicenin-2 hydrocolloid film on Sprague Dawley rats. *BMC Complementary Altern. Med.* **2019**, *19*, 20. [CrossRef]
44. Sunnetha, B.V.; Chiranjeevi; Sreenivasulu; Jayanthi; Venugopal; Akanksha; Naik, N.; Sravani; Reddy, P.K.; Raju, S. Formulation and evaluation of aloe vera herbal ointment [anti-inflammatory anti-oxidant activity]. *World J. Pharm. Res.* **2019**, *8*, 688–699.
45. Saleem, M.; Nazir, M.; Hussain, H.; Tousif, M.I.; Elsebai, M.F.; Riaz, N.; Akhtar, N. Natural Phenolics as Inhibitors of the Human Neutrophil Elastase (HNE) Release: An Overview of Natural Anti-inflammatory Discoveries during Recent Years. *Antiinflamm. Antiallergy Agents Med. Chem.* **2018**, *17*. [CrossRef]
46. Husen, S.A.; Setyawan, M.F..; Syadzha, M.F.; Susilo, R.J.K.; Hayaza, S.; Ansori, A.N.M.; Alamsjah, M.A.; Ilmi, Z.N.; Wulandari, P.A.C.; Pudjiastuti, P.; et al. A Novel Therapeutic Effects of *Sargassum ilicifolium* Alginate and Okra (*Abelmoschus esculentus*) Pods Extracts on Open Wound Healing Process in Diabetic Mice. *Res. J. Pharm. Technol.* **2020**, *13*, 2764–2770.
47. Maruyama, K.; Asai, J.; Ii, M.; Thorne, T.; Losordo, D.W.; D'Amore, P.A. Decreased macrophage number and activation leads to reduced lymphatic vessel formation and contributes to impaired diabetes wound healing. *Am. J. Pathol.* **2007**, *170*, 1178–1191. [CrossRef]
48. Reilkoff, R.A.; Bucala, R.; Herzog, E.L. Fibrocytes: Emerging effector cells in chronic inflammation. *Nat. Rev. Immunol.* **2011**, *11*, 427–435. [CrossRef] [PubMed]
49. Portou, M.J.; Baker, D.; Abraham, D.; Tsui, J. The innate immune system, toll-like receptors and dermal wound healing: A review. *Vasc. Pharm.* **2015**, *71*, 31–36. [CrossRef]
50. Wild, T.; Rahbarnia, A.; Kellner, M.; Sobotka, L.; Eberlein, T. Basics innutrition and wound healing. *Nutrition* **2010**, *26*, 862–866. [CrossRef]
51. DiPietro, L.A. Wound healing: The role of the macrophage and other immune cells. *Shock* **1995**, *4*, 233–240. [CrossRef]
52. Heublein, H.; Bader, A.; Giri, S. Preclinical and clinical evidence for stem cell therapies as treatment for diabetic wounds. *Drug Discov. Today* **2015**, *20*, 703–717. [CrossRef] [PubMed]

Article

Chemical Investigation of Diketopiperazines and N-Phenethylacetamide Isolated from *Aquimarina* sp. MC085 and Their Effect on TGF-β-Induced Epithelial–Mesenchymal Transition

Myong Jin Lee [1,†], Geum Jin Kim [2,3,†], Myoung-Sook Shin [1], Jimin Moon [2], Sungjin Kim [1], Joo-Won Nam [2], Ki Sung Kang [1,*] and Hyukjae Choi [2,3,*]

1 College of Korean Medicine, Gachon University, Seongnam 13120, Korea; myongene@naver.com (M.J.L.); ms.shin@gachon.ac.kr (M.-S.S.); qkrnsld@naver.com (S.K.)
2 College of Pharmacy, Yeungnam University, Gyeongsan 38541, Korea; kimgeumjin@naver.com (G.J.K.); hyp1112@yu.ac.kr (J.M.); jwnam@yu.ac.kr (J.-W.N.)
3 Research Institute of Cell Culture, Yeungnam University, Gyeongsan 38541, Korea
* Correspondence: kkang@gachon.ac.kr (K.S.K.); h5choi@yu.ac.kr (H.C.); Tel.: +82-31-750-5402 (K.S.K.); +82-53-810-2824 (H.C.)
† These authors contributed equally to this work.

Abstract: Chemical investigations of *Aquimarina* sp. MC085, which suppressed TGF-β-induced epithelial–mesenchymal transition (EMT) in A549 human lung cancer cells, led to the isolation of compounds **1–3**. Structural characterization using spectroscopic data analyses in combination with Marfey's analysis revealed that they were two diketopiperazines [*cyclo*(L-Pro-L-Leu) (**1**) and *cyclo*(L-Pro-L-Ile) (**2**)] and one N-phenethylacetamide (**3**). *Cyclo*(L-Pro-L-Leu) (**1**) and N-phenethylactamide (**3**) inhibited the TGF-β/Smad pathway and suppressed the metastasis of A549 cells by affecting TGF-β-induced EMT. However, *cyclo*(L-Pro-L-Ile) (**2**) downregulated mesenchymal factors via a non-Smad-mediated signaling pathway.

Keywords: *Aquimarina* sp.; diketopiperazine; N-phenethylacetamide; epithelial-mesenchymal transition (EMT); A549 cells

Citation: Lee, M.J.; Kim, G.J.; Shin, M.-S.; Moon, J.; Kim, S.; Nam, J.-W.; Kang, K.S.; Choi, H. Chemical Investigation of Diketopiperazines and N-Phenethylacetamide Isolated from *Aquimarina* sp. MC085 and Their Effect on TGF-β-Induced Epithelial–Mesenchymal Transition. *Appl. Sci.* **2021**, *11*, 8866. https://doi.org/10.3390/app11198866

Academic Editor: Panagiota Diamantopoulou

Received: 6 August 2021
Accepted: 18 September 2021
Published: 23 September 2021

Publisher's Note: MDPI stays neutral with regard to jurisdictional claims in published maps and institutional affiliations.

1. Introduction

Marine natural products are considered as rich sources of cytotoxins, many of which could serve as lead compounds for anticancer drug development [1]. Recently, many marine natural products have become accepted as the biosynthetic products of marine microorganisms [2]. Chemical investigations on marine microorganisms have resulted in the discovery of anticancer agents, such as bretuximab vedotin and salinosporamide A [3].

Epithelial–mesenchymal transition (EMT) is an important morphological process that differentiates epithelial cells into mesenchymal phenotype. It is characterized by the loss of polarity, cell–cell contact, and gain of mesenchymal markers. Epithelial cadherin is degraded in the plasma membrane, and desmosomes are inhibited by transcription. EMT facilitates the migration and invasion of cancer cells [4–6]. Therefore, EMT inhibitors are effective in cancer chemotherapy, particularly against metastasis.

For many years, plant-derived natural products such as arctigenin [4], betanin [5], camosol [6], curcumin [7], paeoniflorin [8], and tannic acid [9] have been known to regulate EMT-related signaling pathways. However, a few marine derived natural products were also reported to have EMT-modulating activities. Eribulin mesylate, a marine-derived natural product and FDA-approved anticancer drug, is an EMT inhibitor [10]. In addition, marine microbial natural products such as biemamides [11], actinomycin V [12], pentabromopseudilin [13], and androsamide [14] were shown to target the EMT signaling pathway.

As part of our efforts to discover bioactive natural products from marine microorganisms, we found that the ethyl acetate extract of marine-derived *Aquimarina* sp. MC085 inhibited TGF-β-induced EMT in A549 cells. The genus *Aquimarina* belongs to the family *Flavobacteriaceae* and is known for its aerobic and halophilic characteristics. Chemical investigations of this genus were rarely reported. The *Aquimarina* genus was reported to produce algicidal proteins against toxic cyanobacterium *Microcystis aeruginosa* [15]. Genome mining of *Aquimarina* sp. Ap349 resulted in the discovery of cuniculenes 6A and 6 B derived from polyketide synthase [16]. However, no other natural product with biological activity was reported from the bacterium in this genus. Bioactivity-guided isolation and chemical investigations of *Aquimarina* sp. MC085 led to the isolation of three natural products, and a spectroscopic data analysis led to the structural assignment of the isolates.

2. Materials and Methods

2.1. General Experimental Procedure

Optical rotation was measured using a Jasco DIP-1000 polarimeter (Tokyo, Japan). Nuclear magnetic resonance (NMR) spectra were recorded using a 250 MHz Bruker NMR spectrometer (DMX 250) and 600 MHz Varian NMR spectrometer (VNS-600, Palo Alto, Santa Clara, CA, USA) at the Core Research Support Center for Natural Products and Medical Materials (CRCNM). Low-resolution electrospray ionization MS (LR–ESI–MS) was performed using an Agilent 6120 single-quadrupole mass spectrometer (Agilent Technologies, Santa Clara, CA, USA) with a C3 column (Agilent SB-C3 Zorbax, 5 μm, 4.6 × 150 mm). Isolation of the compounds was carried out using a Waters 1525 binary high-performance liquid chromatography (HPLC) pump having a Waters 996 photodiode array (PDA) with a reversed-phase HPLC (RS Tech, Cheongju, Republic of Korea, Hector-M 5 μm C18, 250 × 4.6 mm).

2.2. Fermentation of *Aquimarina sp.* MC085 and Preparation of Extracts

The bacterial strain *Aquimarina* sp. MC085 (GenBank Accession No. MG016025) was incubated at 25 °C in a shaking incubator at 150 rpm in 5 L of SYP media. After 7 days, the broth media of strain MC085 was extracted twice with ethyl acetate and the combined extract was evaporated.

2.3. Acid Hydrolysis and C3 Marfey's Analysis of Compounds **1** and **2**

Each compound (100 μg) was dissolved in 6 N HCl (400 μL) in 4 mL glass vials. The vials were incubated at 110 °C for 4 h with stirring. The resulting acid hydrolysates were dried under a stream of N_2 gas. The dried acid hydrolysates were resuspended in deionized water. After the removal of water under a N_2 gas stream, the hydrolysates were dissolved in 40 μL of MeOH and treated with 60 μL of 1.0 M $NaHCO_3$ and 25 μL of 1% L-FDAA (1-fluoro-2,4-dinitrophenyl-5-L-alanine amide) in acetone. The reaction mixtures in each vial were incubated at 40 °C for 1 h with stirring, and the reaction was quenched by the addition of 60 μL of 1 N HCl. The resulting solutions (10 μL) were injected to LC-ESI-MS; Agilent Zorbax SB-C3 column, 5 μm, 150 × 4.6 mm, 50 °C, 1.0 mL/min, solvent A (deionized water), solvent B [95% MeOH + 5% of (95% CH3CN + 5% formic acid)], A:B = 70:30→45:55 (95 min)→0:100 (97 min)→0:100 (104 min)→70:30 (105 min)→70:30 (110 min). The reaction products with FDAA-derivatized hydrolysates were analyzed in the negative mode of the LC-MS system. The retention time of the DAA-derivatives of constituting amino acids was compared with that of the authentic standards (DAA-L-Pro: 15.5 min, DAA-D-Pro: 19.0 min, DAA-L-*allo*-Ile: 39.5 min, DAA-L-Ile: 41.8 min, DAA-L-Leu: 42.6 min, DAA-D-*allo*-Ile: 61.7 min, DAA-D-Ile: 64.5 min, DAA-D-Leu: 65.0 min).

2.4. Cell Culture and Treatment

A549 (human lung carcinoma epithelial) cells were incubated in RPMI 1640 medium supplemented with 10% (*v*/*v*) fetal bovine serum and 1% penicillin–streptomycin at 37 °C in a humidified atmosphere of 5% CO_2 and 95% air. The cells were then treated with compounds **1–3**, respectively, in medium for 6 h. This was followed by treatment with TGF-β (5 ng/mL) for 42 h in the presence or absence of compounds at the indicated concentrations.

2.5. Cell Viability Assay

An EZ-Cytox assay kit (DoGenBio, Seoul, Korea) was used to assess cell viability. Cells (5×10^4 cells/well) were seeded into a 96-well plate. After 24 h, compounds **1–3** were added to serum-free medium for 6 h before TGF-β (5 ng/mL) stimulation. EZ-Cytox reagent was added to the plate and incubated for 1 h at 37 °C. The absorbance at 450 nm was measured using an enzyme-linked immunosorbent assay (ELISA) plate reader.

2.6. Western Blotting Analysis

A549 cells were treated with compounds **1–3**, as previously described. After 6 h, cells were treated with 5 ng/mL of TGF-β with or without each peptide for 42 h. Cells were lysed in RIPA buffer (50 mM Tris-HCl pH 7.4, 150 mM NaCl, 0.25% deoxycholate, 1% NP-40, and 1 mM EDTA), and protein concentration was determined using BCA reagent (ThermoScientific, Rockford, IL, USA) according to the manufacturer's instructions.

Proteins were separated by sodium dodecyl sulfate polyacrylamide gel electrophoresis (SDS-PAGE) and then transferred to polyvinylidene difluoride (PVDF) membranes (Millipore, Burlington, MA, USA). After the transfer, membranes were blocked with 5% nonfat dried milk/TBST (20 mM Tris pH 7.4, 150 mM NaCl, and 0.2% Tween-20) for 1 h and then incubated overnight at 4 °C with primary antibody. The antibodies used were anti-Smad2/3, phosphorylated Smad2/3, N-cadherin, E-cadherin, Vimentin, Snail, β-catenin, and GAPDH from Cell Signaling Technology (Danvers, MA, USA). After the wash with TBST, the membranes were probed with HRP-labeled secondary antibodies. Finally, an enhanced chemiluminescence detection kit (Fisher Scientific, Rockford, IL, USA) was used to visualize the immunoreactive proteins.

2.7. Gelatin Zymography

Cell supernatants of serum-free cultures were concentrated using Amicon Ultra-4 Centrifugal Filter Devices (Millipore, Billerica, MA, USA). Matrix Metalloproteinase-2 (MMP-2) activity was determined using a Zymogram–PAGE System (Komabiotech, Seoul, Korea) according to the manufacturer's instructions. In brief, after electrophoresis, the gel was washed for 30 min to remove the SDS and then incubated overnight at 37 °C with a developing buffer. Then, the gel was stained with Coomassie Blue R-250 and distained with methanol and acetic acid. The gelatinolytic activities were then visualized as a white band and the gels were scanned.

2.8. Statistical Analysis

Results are expressed as mean ± standard deviations (SD). The statistical significance between test groups was determined using a one-way ANOVA multiple comparison test with Turkey's post-test using GraPhad Prism 5 software (GraphPad Software Inc., San Diego, CA, USA). $p < 0.05$ was considered statistically significant.

3. Results

3.1. Structural Identification of Compounds 1–3

Compound **1** was obtained as a white powder, {$[\alpha]^{22}_D$ −54.47° (*c* 0.14, CH_3OH)}. The ^{13}C NMR spectrum of **1** in DMSO-*d6* showed distinctive 11 carbon signals including two amide carbonyls at δ_C 170.32 and 166.51, three sp3 carbons with the attachment of N atom at δ_C 58.44, 52.59, and 44.83, and six sp3 carbons without N attachment at δ_C 37.76, 27.40, 24.04, 22.79, 22.44, and 21.88. The ^1H NMR spectrum of **1** in DMSO-*d6* showed a broad NH proton signal at δ_H 8.00, signals for two α protons at δ_H 4.19, and 4.00, N-attached methylene protons at δ_H 3.42–3.30, two doublet methyl signals at δ_H 0.87 (J = 6.3 Hz) and 0.86 (J = 6.3 Hz), along with the residual seven protons at δ_H 2.22–1.30. The ESI-MS spectrum (positive ion detection mode) showed a peak corresponding to a protonated molecule at m/z 211.4 $[M + H]^+$. These data indicated that the molecular formula of compound **1** was $C_{11}H_{18}N_2O_2$. The 2D NMR data analysis led to the planar structure of **1** as a cyclic dipeptide, *cyclo*(Pro-Leu), as shown in Figure 1. The absolute configuration of **1** was determined to be *cyclo*(L-Pro-L-Leu) using C3 Marfey's method (Figure S8) [17].

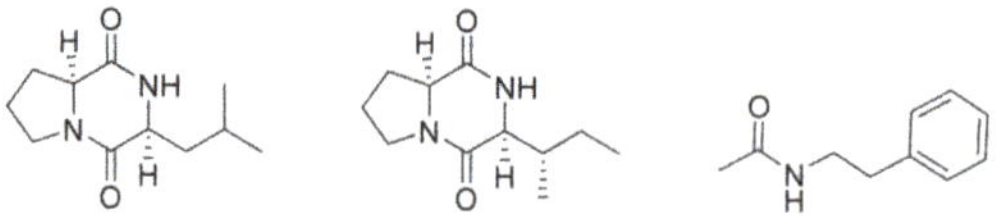

Figure 1. Structures of compounds **1–3** from *Aquimarina* sp. MC085.

Compound **2** {$[\alpha]^{22}_D$ −48.65° (*c* 0.07, CH_3OH)} showed a protonated ion at m/z 211.4 $[M + H]^+$ in the positive ion mode of the ESI-MS spectrum. The ^1H NMR spectrum of **2** showed structural features similar to those of compound **1**. However, compound **2** showed a distinctive methyl triplet and a methyl doublet, whereas compound **1** showed two methyl doublets. Based on a careful comparison of spectroscopic data, including ^1H and ^{13}C NMR data, compound **2** was identified as *cyclo*(Pro-Ile). In addition, C3 Marfey's analyses configured both constituting amino acids as L (Figure S9) [17].

Compound **3** was obtained as a white powder. Its molecular formula was assigned as $C_{10}H_{13}NO$ based on 1D NMR spectra and the presence of a protonated ion at m/z 164.2 $[M + H]^+$ in the ESI-MS spectrum. The ^1H and ^{13}C NMR spectra of compound **3** in $CDCl_3$ showed five proton signals in the ranges of δ_H 7.15–7.35, and resonances for four sp2 methine carbons at δ_C 138.97, 128.80, 128.69, and 126.57, corresponding to a mono-substituted phenyl ring. Moreover, signals of two methylene units (δ_C 40.78, δ_H 2.80; δ_C 35.68, δ_H 3.49) and an acetyl unit (δ_C 170.35; δ_C 23.30, δ_H 1.92) were observed. By comparing the spectroscopic data with the values provided in the literature, the structure of compound **3** was confirmed to be N-phenethylacetamide [18]. Compounds **1–3** were isolated from microorganisms of the genus *Aquimarina* for the first time.

3.2. The Cell Viability Effect on Comparison of Compounds 1–3 on TGF-β-Induced EMT of A549 Cells

First, we investigated the cytotoxic effects of compounds **1–3** with or without TGF-β on A549 cells using an EZ-Cytox assay. As presented in Figure 2, the cells were treated with various concentrations of compounds **1–3** (100, 50, 25, and 12.5 μM) for 24 h. The cell viability analysis showed that none of the components affected cell viability at concentrations below 50 μM; however, compound **2** showed a significant decrease cell viability at 100 μM concentration. Therefore, further assays were conducted at a concentration of 50 μM.

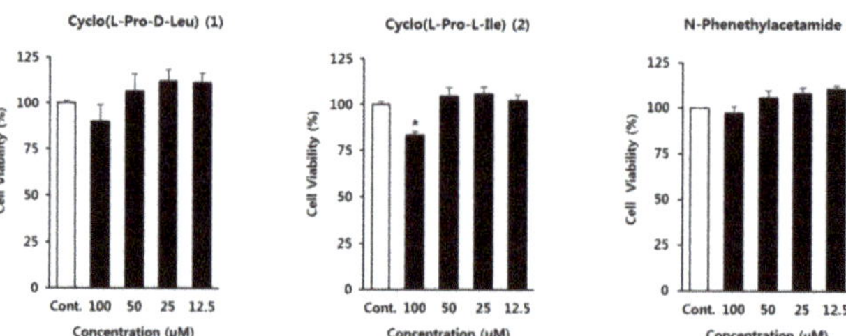

Figure 2. Comparison of the effects of compounds **1–3** on TGF-β-induced epithelial to mesenchymal transition (EMT) of A549 cells. The cells were treated with different concentrations of **1–3** before TGF-β treatment for 24 h and viability was determined using the EZ-Cytox assay. The data represent the means ± S.D. of three independent experiments. * $p < 0.05$ compared with the control.

3.3. *Comparison of the Protein Expression Profiles of A549 Cells to Determine the Effect of the Treatment with Compounds* **1–3** *(50 µM), via the Smad2/3 Signaling Pathway, on the TGF-β-Mediated EMT of the Cells*

To observe the expression levels of proteins associated with canonical Smad2/3, A549 cells were treated with 50 µM of the test compounds, followed by a TGF-β treatment. Phosphorylation of Smad2/3 was measured using western blotting. On comparison with the control group, the results showed that TGF-β induced the phosphorylation of Smad2/3. Phosphorylated Smad2/3 was inhibited by compounds **1** and **3** (Figure 3A). We also examined the effect of compounds **1–3** on TGF-β-mediated expression of EMT-related markers in A549 cells. N-Phenethylacetamide (**3**) dramatically downregulated mesenchymal phenotypic markers (N-cadherin, vimentin, and Snail) induced by TGF-β (p = 0.0014). The levels of vimentin and Snail were reduced by compound **2** in the TGF-β-treated A549 cells. However, β-catenin levels were not changed in the compounds **1–3**-treated cells exposed or not exposed to TGF-β (Figure 3B).

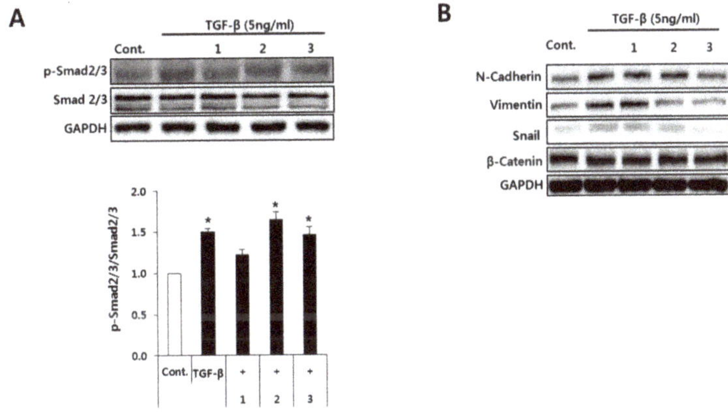

Figure 3. Comparison of the protein expression profiles of A549 cells to determine the effect of the treatment with compounds **1–3** (50 µM, respectively), via the Smad2/3 signaling pathway, on the TGF-β-mediated EMT of the cells. (**A**) Phosphorylation of Smad2/3 was detected by Western blot analysis and densitometry. Error bars denote SD from three independent replicates. (*) indicates statistical significance ($p < 0.05$) when compared to control. (**B**) The levels of TGF-β-induced EMT-related markers were determined by Western blotting.

3.4. *Comparison of the Effects of the Compounds* **1–3** *(50 µM) on the Activity of MMP-2 in A549 Human Lung Carcinoma Cells, Tested Using Gelatin Zymography*

MMP-2 is the main proteolytic collagenase among MMPs and plays a major role in the invasion and metastasis of cancer cells. Therefore, we investigated the effects of compounds **1–3** on MMP-2 activity. The 62-KDa form was identified as MMP-2. MMP-2 expression was increased by TGF-β stimulation. All cells treated with compounds **1–3** showed decreased MMP-2 activity compared to cells subjected to TGF-β treatment alone (Figure 4).

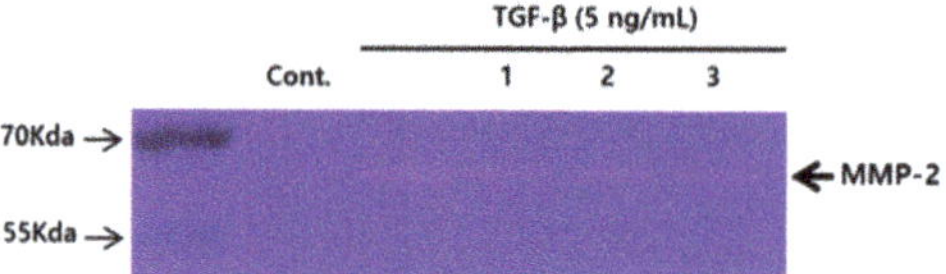

Figure 4. Comparison of the effects of compounds **1–3** (50 μM) on matrix metalloproteinase-2 (MMP-2) activity in A549 human lung carcinoma cells, tested using gelatin zymography. The cell supernatants were harvested and concentrated by centrifugation using Amicon Ultra-4 centrifugal filter devices. Clear bands represent gelatin-cleaved areas.

4. Discussion

EMT is induced by several signaling pathways, including transforming growth factor-β1 (TGF-β1), hypoxia, notch, and hedgehog [19]. In our study, EMT of human lung cancer A549 cells was induced by TGF- β. EMT is characterized by the loss of epithelial markers, including vimentin, N-cadherin, fibronectin, and α-smooth muscle actin, and an increase in mesenchymal markers [20]. Therefore, we analyzed the protein expression of mesenchymal markers by immunoblotting. MMP-2 is a proteolytic enzyme that induces EMT by breaking down the extracellular matrix of malignant cells [21]. In this study, MMP-2 activity was evaluated using gelatin zymography.

Compound **1** is known to have diverse bioactivities, including antibacterial [22], antifungal [23], antifouling [24], anti-biofilm activities [25], quorum sensing modulation [26,27], inhibition of dengue virus replication [28], and induction of mitochondria-mediated apoptosis and tumor growth suppression [29]. Compound **2** was reported to have antifungal [30] and quorum sensing inhibitory activities [26]. Compound **3** was reported to inhibit chymotrypsin A [31], increase the lifespan of *Caenorhabditis elegans* by SIR-2.1 induction [32], and reverse the resistance of doxorubicin-resistant leukemia P388 cells [33]. However, the EMT-related bioactivities of compounds **1–3** were not previously investigated. Compounds **1** and **3** downregulated MMP-2 through the inhibition of TGF-β-mediated phosphorylation of Smad2/3 in A549 cells. Compound **3** also reversed TGF-β-induced EMT markers, N-cadherin, vimentin, and Snail. Although compound **2** exhibited activities through a non-Smad-mediated signaling pathway, it attenuated TGF-β-dependent induction of mesenchymal markers, such as vimentin and Snail. In addition, it modulated the expression level of MMP-2 compared to that in the untreated controls in the A549 cell line. High MMP expression is associated with tumor invasion and metastasis. The present study showed that the increased expression of MMP-2 by TGF-β treatment was inhibited by compounds **1–3** (Figure 5). Smad2/3 activated by TGF-β induction translocates to the nucleus, and β-catenin might form a complex with active Smad2/3. The complex may interact with various transcription factors and regulate EMT-related genes. *Cyclo*(L-Pro-L-Leu) (**1**) and N-phenethylactamide (**3**) inhibited TGF-β/Smad pathways and suppressed metastasis of A549 cells by affecting TGF-β-induced EMT. However, *cyclo*(L-Pro-L-Ile) (**2**) downregulated mesenchymal factors via a non-Smad-mediated signaling pathway. The chemical investigation of *Aquimarina* sp. MC085 led to the identification of two diketopiperazines [*cyclo*(L-Pro-L-Leu) (**1**) and *cyclo*(L-Pro-L-Ile) (**2**)] and one N-phenethylacetamide (**3**), which inhibit TGF-β-mediated EMT.

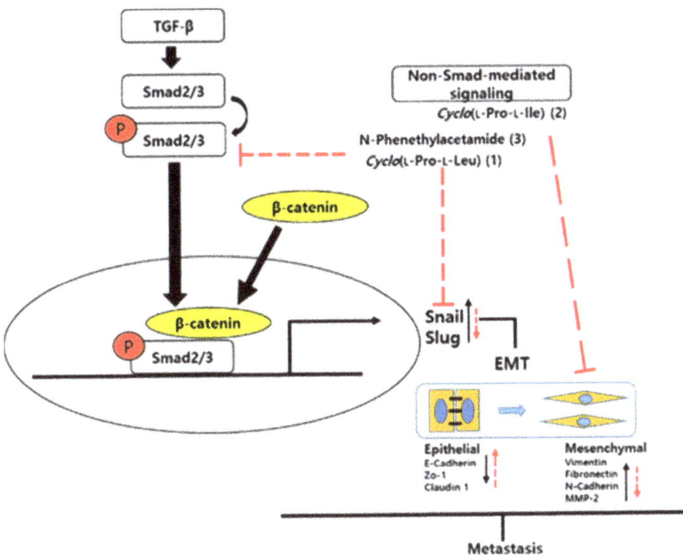

Figure 5. Proposed mechanism of action of compounds **1–3** on TGF-β-induced EMT of A549 cells.

5. Conclusions

The chemical investigation of *Aquimarina* sp. MC085 led to the isolation of compounds **1–3**, which suppressed TGF-β-induced EMT of A549 human lung cancer cells. *Cyclo*(L-Pro-L-Leu) (**1**) and N-phenethylactamide (**3**) inhibited TGF-β/Smad pathways and suppressed the metastasis of A549 cells by affecting TGF-β-induced EMT. However, *cyclo* (L-Pro-L-Ile) (**2**) downregulated mesenchymal factors via a non-Smad-mediated signaling pathway.

Supplementary Materials: The following are available online at https://www.mdpi.com/article/10.3390/app11198866/s1, Figure S1: LR-ESI-MS data of **1**; Figure S2: ^{1}H NMR spectrum (250 MHz) of **1** in DMSO-d_6; Figure S3: ^{13}C NMR (63 MHz) spectrum of **1** in DMSO-d_6; Figure S4: HMQC spectrum of **1** in DMSO-d_6; Figure S5: HMBC spectrum of **1** in DMSO-d_6; Figure S6: ^{1}H NMR spectrum (600 MHz) of **1** in CD_3OD; Figure S7: ^{13}C NMR (150 MHz) spectrum of **1** in CD_3OD; Figure S8: C3 Marfey's analysis of **1**; Figure S9: LR-ESI-MS data of **2**; Figure S10: ^{1}H NMR spectrum (250 MHz) of **2** in CD_3OD; Figure S11: ^{3}C NMR (63 MHz) spectrum of **2** in CD_3OD; Figure S12: C3 Marfey's analysis of **2**; Figure S13: LR-ESI-MS data of **3**; Figure S14: ^{1}H NMR spectrum (250 MHz) of **3** in $CDCl_3$; Figure S15: ^{13}C NMR (63 MHz) spectrum of **3** in $CDCl_3$.

Author Contributions: Conceptualization, K.S.K. and H.C.; formal analysis, M.J.L., G.J.K., M.-S.S., J.M. and S.K.; investigation, M.J.L. and J.-W.N.; writing—original draft preparation, M.J.L., G.J.K. and H.C.; writing—review and editing, K.S.K.; visualization, M.J.L. and G.J.K.; supervision, K.S.K. and H.C.; project administration, K.S.K. and H.C.; funding acquisition, K.S.K. and H.C. All authors have read and agreed to the published version of the manuscript.

Funding: This research was supported by the 2020 Yeungnam University Research Grants and the research grants from the Basic Science Research Program through the National Research Foundation of Korea (NRF) (grant number: 2019R1F1A1059173 and 2020R1A6A1A03044512).

Institutional Review Board Statement: Not applicable.

Informed Consent Statement: Not applicable.

Data Availability Statement: The data that support the findings of this study are available from the corresponding author upon reasonable request.

Conflicts of Interest: The authors declare no conflict of interest.

References

1. Newman, D.J.; Cragg, G.M. Natural products as sources of new drugs over the nearly four decades from 01/1981 to 09/2019. *J. Nat. Prod.* **2020**, *83*, 770–803. [CrossRef] [PubMed]
2. Gerwick, W.H.; Moore, B.S. Lessons from the past and charting the future of marine natural products drug discovery and chemical biology. *Chem. Biol.* **2012**, *19*, 85–98. [CrossRef] [PubMed]
3. Jiménez, C. Marine Natural Products in Medicinal Chemistry. *ACS Med. Chem. Lett.* **2018**, *9*, 959–961. [CrossRef] [PubMed]
4. Xu, Y.; Lou, Z.; Lee, S.-H. Arctigenin represses TGF-β-induced epithelial mesenchymal transition in human lung cancer cells. *Biochem. Biophys. Res. Commun.* **2017**, *493*, 934–939. [CrossRef]
5. Sutariya, B.; Saraf, M. Betanin, isolated from fruits of Opuntia elatior Mill attenuates renal fibrosis in diabetic rats through regulating oxidative stress and TGF-β pathway. *J. Ethnopharmacol.* **2017**, *198*, 432–443. [CrossRef]
6. Giacomelli, C.; Daniele, S.; Natali, L.; Iofrida, C.; Flamini, G.; Braca, A.; Trincavelli, M.L.; Martini, C. Carnosol controls the human glioblastoma stemness features through the epithelial-mesenchymal transition modulation and the induction of cancer stem cell apoptosis. *Sci. Rep.* **2017**, *7*, 15174. [CrossRef]
7. Li, W.; Jiang, Z.; Xiao, X.; Wang, Z.; Wu, Z.; Ma, Q.; Cao, L. Curcumin inhibits superoxide dismutase-induced epithelial-to-mesenchymal transition via the PI3K/Akt/NF-κB pathway in pancreatic cancer cells. *Int. J. Oncol.* **2018**, *52*, 1593–1602. [CrossRef]
8. Wang, Z.; Liu, Z.; Yu, G.; Nie, X.; Jia, W.; Liu, R.-E.; Xu, R. Paeoniflorin inhibits migration and invasion of human glioblastoma cells via suppression transforming growth factor β-induced epithelial–mesenchymal transition. *Neurochem. Res.* **2018**, *43*, 760–774. [CrossRef]
9. Pattarayan, D.; Sivanantham, A.; Krishnaswami, V.; Loganathan, L.; Palanichamy, R.; Natesan, S.; Muthusamy, K.; Rajasekaran, S. Tannic acid attenuates TGF-β1-induced epithelial-to-mesenchymal transition by effectively intervening TGF-β signaling in lung epithelial cells. *J. Cell. Physiol.* **2018**, *233*, 2513–2525. [CrossRef]
10. Dybdal-Hargreaves, N.F.; Risinger, A.L.; Mooberry, S.L. Eribulin mesylate: Mechanism of action of a unique microtubule-targeting agent. *Clin. Cancer Res.* **2015**, *21*, 2445–2452. [CrossRef]
11. Zhang, F.; Braun, D.R.; Ananiev, G.E.; Hoffmann, F.M.; Tsai, I.-W.; Rajski, S.R.; Bugni, T.S. Biemamides A–E, inhibitors of the TGF-β pathway that block the epithelial to mesenchymal transition. *Org. Lett.* **2018**, *20*, 5529–5532. [CrossRef] [PubMed]
12. Lin, S.; Zhang, C.; Liu, F.; Ma, J.; Jia, F.; Han, Z.; Xie, W.; Li, X. Actinomycin V inhibits migration and invasion via suppressing snail/slug-mediated epithelial-mesenchymal transition progression in human breast cancer MDA-MB-231 cells in vitro. *Mar. Drugs* **2019**, *17*, 305. [CrossRef] [PubMed]
13. Shih-Wei, W.; Chih-Ling, C.; Kao, Y.-C.; Martin, R.; Knölker, H.-J.; Shiao, M.-S.; Chen, C.-L. Pentabromopseudilin: A myosin V inhibitor suppresses TGF-β activity by recruiting the type II TGF-β receptor to lysosomal degradation. *J. Enzym. Inhib. Med. Chem.* **2018**, *33*, 920–935. [CrossRef]
14. Lee, J.; Gamage, C.D.; Kim, G.J.; Hillman, P.F.; Lee, C.; Lee, E.Y.; Choi, H.; Kim, H.; Nam, S.-J.; Fenical, W. Androsamide, a cyclic tetrapeptide from a marine *Nocardiopsis* sp., suppresses motility of colorectal cancer cells. *J. Nat. Prod.* **2020**, *83*, 3166–3172. [CrossRef]
15. Chen, W.M.; Sheu, F.S.; Sheu, S.Y. Novel L-amino acid oxidase with algicidal activity against toxic cyanobacterium *Microcystis aeruginosa* synthesized by a bacterium *Aquimarina* sp. *Enzym. Microb. Technol.* **2011**, *49*, 372–379. [CrossRef]
16. Helfrich, E.J.; Ueoka, R.; Dolev, A.; Rust, M.; Meoded, R.A.; Bhushan, A.; Califano, G.; Costa, R.; Gugger, M.; Steinbeck, C. Automated structure prediction of trans-acyltransferase polyketide synthase products. *Nat. Chem. Biol.* **2019**, *15*, 813–821. [CrossRef] [PubMed]
17. Vijayasarathy, S.; Prasad, P.; Fremlin, L.J.; Ratnayake, R.; Salim, A.A.; Khalil, Z.; Capon, R.J. C3 and 2D C3 Marfey's methods for amino acid analysis in natural products. *J. Nat. Prod.* **2016**, *79*, 421–427. [CrossRef]
18. Llinarés, J.; Elguero, J.; Faure, R.; Vincent, E.J. Carbon-13 NMR studies of nitrogen compounds. I—substituent effects of amino, acetamido, diacetamido, ammonium and trimethylammonium groups. *Org. Magn. Reson.* **1980**, *14*, 20–24. [CrossRef]
19. Luo, W.; Liu, Q.; Jiang, N.; Li, M.; Shi, L. Isorhamnetin inhibited migration and invasion via suppression of Akt/ERK-mediated epithelial-to-mesenchymal transition (EMT) in A549 human non-small-cell lung cancer cells. *Biosci. Rep.* **2019**, *39*, BSR20190159. [CrossRef]
20. Chen, K.-J.; Li, Q.; Wen, C.-M.; Duan, Z.-X.; Zhang, J.Y.; Xu, C.; Wang, J.-M. Bleomycin (BLM) induces epithelial-to-mesenchymal transition in cultured A549 cells via the TGF-β/Smad signaling pathway. *J. Cancer* **2016**, *7*, 1557. [CrossRef]
21. Agraval, H.; Yadav, U.C. MMP-2 and MMP-9 mediate cigarette smoke extract-induced epithelial-mesenchymal transition in airway epithelial cells via EGFR/Akt/GSK3β/β-catenin pathway: Amelioration by fisetin. *Chem.-Biol. Interact.* **2019**, *314*, 108846. [CrossRef]
22. Rhee, K.-H. Isolation and characterization of Streptomyces sp. KH-614 producing anti-VRE (vancomycin-resistant enterococci) antibiotics. *J. Gen. Appl. Microbiol.* **2002**, *48*, 321–327. [CrossRef] [PubMed]
23. Rhee, K.-H. Purification and identification of an antifungal agent from *Streptomyces* sp. KH-614 antagonistic to rice blast fungus, *Pyricularia oryzae*. *J. Microbiol. Biotechnol.* **2003**, *13*, 984–988.
24. Li, X.; Dobretsov, S.; Xu, Y.; Xiao, X.; Hung, O.S.; Qian, P.-Y. Antifouling diketopiperazines produced by a deep-sea bacterium, *Streptomyces fungicidicus*. *Biofouling* **2006**, *22*, 187–194. [CrossRef] [PubMed]

25. Gowrishankar, S.; Kamaladevi, A.; Ayyanar, K.S.; Balamurugan, K.; Pandian, S.K. Bacillus amyloliquefaciens-secreted cyclic dipeptide–cyclo (L-leucyl-L-prolyl) inhibits biofilm and virulence production in methicillin-resistant *Staphylococcus aureus*. *RSC Adv.* **2015**, *5*, 95788–95804. [CrossRef]
26. Abed, R.M.; Dobretsov, S.; Al-Fori, M.; Gunasekera, S.P.; Sudesh, K.; Paul, V.J. Quorum-sensing inhibitory compounds from extremophilic microorganisms isolated from a hypersaline cyanobacterial mat. *J. Ind. Microbiol. Biotechnol.* **2013**, *40*, 759–772. [CrossRef] [PubMed]
27. Gu, Q.; Fu, L.; Wang, Y.; Lin, J. Identification and characterization of extracellular cyclic dipeptides as quorum-sensing signal molecules from *Shewanella baltica*, the specific spoilage organism of *Pseudosciaena crocea* during 4 °C storage. *J. Agric. Food Chem.* **2013**, *61*, 11645–11652. [CrossRef]
28. Lin, C.-K.; Wang, Y.-T.; Hung, E.-M.; Yang, Y.-L.; Lee, J.-C.; Sheu, J.-H.; Liaw, C.-C. Butyrolactones and diketopiperazines from marine microbes: Inhibition effects on dengue virus type 2 replication. *Planta Med.* **2017**, *83*, 158–163. [CrossRef]
29. Jinendiran, S.; Teng, W.; Dahms, H.-U.; Liu, W.; Ponnusamy, V.K.; Chiu, C.C.-C.; Kumar, B.D.; Sivakumar, N. Induction of mitochondria-mediated apoptosis and suppression of tumor growth in zebrafish xenograft model by cyclic dipeptides identified from *Exiguobacterium acetylicum*. *Sci. Rep.* **2020**, *10*, 13721. [CrossRef]
30. Lind, H.; Sjögren, J.; Gohil, S.; Kenne, L.; Schnürer, J.; Broberg, A. Antifungal compounds from cultures of dairy propionibacteria type strains. *FEMS Microbiol. Lett.* **2007**, *271*, 310–315. [CrossRef]
31. Powers, J.C.; Baker, B.L.; Brown, J.; Chelm, B.K. Inhibition of chymotrypsin A. alpha. with N-acyl-and N-peptidyl-2-phenylethylamines. Subsite binding free energies. *J. Am. Chem. Soc.* **1974**, *96*, 238–243. [CrossRef] [PubMed]
32. Kim, J.H.; Bang, I.H.; Noh, Y.J.; Kim, D.K.; Bae, E.J.; Hwang, I.H. Metabolites Produced by the Oral Commensal Bacterium *Corynebacterium durum* Extend the Lifespan of *Caenorhabditis elegans* via SIR-2.1 Overexpression. *Int. J. Mol. Sci.* **2020**, *21*, 2212. [CrossRef] [PubMed]
33. Kunimoto, S.; Chin-Zhi, X.; Naganawa, H.; Hamada, M.; Masuda, T.; Tajeuchi, T.; Umezawa, H. Reversal of resistance by N-acetyltyramine or N-acetyl-2-phenylethylamine in doxorubicin-resistant leukemia P388 cells. *J. Antibiot.* **1987**, *40*, 1651–1652. [CrossRef] [PubMed]

MDPI
St. Alban-Anlage 66
4052 Basel
Switzerland
Tel. +41 61 683 77 34
Fax +41 61 302 89 18
www.mdpi.com

Applied Sciences Editorial Office
E-mail: applsci@mdpi.com
www.mdpi.com/journal/applsci

www.ingramcontent.com/pod-product-compliance
Lightning Source LLC
LaVergne TN
LVHW070626170726
843515LV00004B/184

* 9 7 8 3 0 3 6 5 3 3 6 7 4 *